Life Writing Series

Life Writing Series

In the **Life Writing Series**, Wilfrid Laurier University Press publishes life writing and new life-writing criticism in order to promote auto-biographical accounts, diaries, letters and testimonials written and/or told by women and men whose political, literary or philosophical purposes are central to their lives. **Life Writing** features the accounts of ordinary people, written in English, or translated into English from French or the languages of the First Nations or from any of the languages of immigration to Canada. **Life Writing** will also publish original theoretical investigations about life writing, as long as they are not limited to one author or text.

Priority is given to manuscripts that provide access to those voices that have not traditionally had access to the publication process.

Manuscripts of social, cultural and historical interest that are considered for the series, but are not published, are maintained in the **Life Writing Archive** of Wilfrid Laurier University Library.

Series Editor
Marlene Kadar
Humanities Division, York University

Manuscripts to be sent to
Brian Henderson, Director
Wilfrid Laurier University Press
75 University Avenue West
Waterloo, Ontario, Canada N2L 3C5

Chasing the Comet

A SCOTTISH-CANADIAN LIFE

Patricia Koretchuk

Wilfrid Laurier University Press

WLU

We acknowledge the support of the Canada Council for the Arts for our publishing program. We acknowledge the financial support of the Government of Canada through the Book Publishing Industry Development Program for our publishing activities.

National Library of Canada Cataloguing in Publication Data

Koretchuk, Patricia, 1934–
    Chasing the comet : a Scottish-Canadian life

(Life writing series)
ISBN 0-88920-407-1

1. Caldow, David.   2. Colony Farm—Biography.   3. Farm managers—British Columbia—Biography. 4. Scottish Canadians—British Columbia—Biography. 5. British Columbia—Biography. I. Caldow, David. II. Title. III. Series.

S417.C24K67 2002   630'.92   C2002-900014-9

© 2002 Patricia Koretchuk

Cover design by Leslie Macredie. Front cover image from Gary W. Kronk's Comets & Meteor Showers Web site <http://comets.amsmeteors.org/>. Original painting by Eric S. Young. With permission. Plaid on spine is Hunting MacMillan courtesy of Gaelic College Foundation.

"Ich lebe mein Leben . . . / I live my life in widening," from Rilke's *Book of Hours: Love Poems to God* by Rainer Maria Rilke, translated by Anita Barrows and Joanna Macy © 1996 by Anita Barrows and Joanna Macy. Used by permission of Riverhead Books, a division of Penguin Putnam, Inc.

Wilfrid Laurier University Press
Waterloo, Ontario, Canada   N2L 3C5
www.wlupress.wlu.ca

Printed in Canada

I live my life in widening circles
that reach out across the world.
I may not complete this last one
but I give myself to it.

I circle around God, around the primordial tower.
I've been circling for thousands of years
and I still don't know: am I a falcon,
a storm, or a great song?

—Rainer Maria Rilke

He wasn't a member of
our University Womens'
Club, but he would
have added some humour
and substance to our
discussions. Enjoy!

Patricia Koretchuk

TABLE OF CONTENTS

How this Story Came to Be

This book began with a request from one of my husband's best friends, George Caldow, who asked me to write his father's story, intending it to be a gift for his father's ninetieth birthday. Because George had enjoyed and shared many of his father's memories over the years, he wanted them recorded so they could be savoured by the family, perhaps through the generations. David Caldow was eighty-nine at the time.

When he was told of George's idea, David thought it was amusing but he also thought it would be a good idea for his grandchildren and later generations to know their family story. When we began, the Caldow family was my only intended audience. For the writing, I agreed to be paid expenses—enough to cover the cost of paper and ink cartridges. In return, I asked that David would give me publishing rights. I intended to write something of about three thousand words, like other life stories I had written and published in periodicals—something to perhaps become part of a collection of immigrant stories.

To be honest, when I accepted the task, I wasn't sure that David and I could collaborate successfully because of our many differences. The most obvious one is that he is a man and I am a woman, and we all know what difficulties that difference can create. Also, he is farm folk and I am city folk. His brogue identifies him as immigrant Scottish, while I am first generation English-

Canadian. He had matured at the end of the Victorian era, while I had attended Simon Fraser University during the 1960s. He had lived a long time, happy and successful on Colony Farm (located at Essondale Provincial Mental Hospital), which I remembered from the perspective of a fear-filled teen, trembling at the sight of the place when visiting my severely depressed mother over a seven-month period. David remembers pastoral farm scenes surrounding the hospital grounds, while I remember the sight of barred windows and the clanking sound made when keys turned in iron-clad doors that locked my mother away from me. Nevertheless, I accepted George Caldow's request to write his father's story, partly because I was curious.

I was intrigued by the possibility of learning our friend's history. I knew David's memories of his work at the now-dismantled and transformed Colony Farm and Tranquille would provide interesting historical details that would otherwise be lost from British Columbia's history. Also, personally, I thought that researching the mental hospital might exorcise the ghostly sounds I carried within, although even thinking of doing so was disquieting. However, previous successful writing experiences gave me the confidence I needed to attempt the greater challenge posed by David Caldow's story.

I see myself as an artist creating true-to-life portraits, using words instead of paint. Using a hanging file instead of a palette, I collect details, vignettes, colourful words, and idiomatic phrases, then organize them into a narrative of the person's lifetime. Interviews and research develop contrasts and shades of meaning, highlight salient features, and reveal the unique theme guiding the choices that built the life. At least, this is my goal. Because I work with living subjects, I know I've succeeded when the person whose life I'm portraying says "Yes, that's exactly the way it was."

David Caldow and I had been acquaintances for many years (about twenty-three). His son George had attended UBC with my husband and had frequently reminisced about Tranquille. The first time we met George's father, he had welcomed us to his home in Abbotsford, a visit prompted by his offer to give us our first dog, a black and white sheepdog named Jiggs. Over the years, we had met and talked casually numbers of times: during Christmas and birthday get-togethers, and also in the Pacific National Exhibition barns (where David supervised the exhibits), and during summers when my children were still young enough to be thrilled by sheep and hogs. Later, as guests at a welcome-home gathering, we had briefly enjoyed David and Peggy's stories of life in Tanzania. But our long acquaintance didn't stop me from being anxious as I drove the five minutes from my White Rock home to his apartment for our first interview. Though I had gladly accepted the risk of writing for a best friend, my acceptance was not given without trepidation.

Thoughts of our past experiences flooded my mind as I made the drive. I stepped along the dirt path between the tall evergreens that sheltered David's ground-level, sliding-screen patio door. I knew this first session would decide the fate of his story. I wanted it to be a creative documentary, a true story, though it would employ elements such as dialogue and character development. I hoped to write a story giving insight into the personal and social values guiding his life, not just a collection of anecdotal memoirs. As I walked past his tiny salad garden bordering one edge of the path near his door, I wondered how this tough-minded, sometimes irascible old Scot would react when I asked him necessary but tough-to-answer personal questions, needed to give his story integrity. As I approached, I heard David's voice.

"Come in. Come in," he said as he slid the screen door aside and beckoned me forward.

As I entered his apartment, it became obvious that this man did not fit the stereotypical image I had of a farmer (perhaps implanted by the song describing a man with a "straw hat and old dirty hanky, mopping [his] face like shoe"). I was struck by David's tidiness and the fact that his rooms still held small, feminine traces of Peggy's life, though she had been dead for several years. Moments later, I was relieved as David continued to destroy my preconceptions.

As he settled me into a comfortable chair near an electrical outlet for my laptop, he asked, "Canna get ye a cup 'o coffee?"

I nodded my "yes" and he returned shortly, carrying a tray with not only coffee in cups on saucers, but also serviettes, sliced cheese, and scones (which he had baked).

"I bake cookies too," he said and spoke enthusiastically of how he loved to cook.

So it was, during our first few minutes together, I learned David Caldow was not a man with stereotypical, early twentieth-century male views of women, relationships, or housework.

As we sipped coffee together, we also got down to business. I explained the procedures that had worked for me before. We would meet every two weeks, at first. Each interview would take roughly one hour. Each time we met, I would give him a printout of the notes of the previous meeting (or of the text in progress) so he could check the writing for accuracy. I told him to write his comments right on the paper because these would be working copies. As I accumulated more information and began writing chapter-length pieces, he would do the same for them. Later we would meet less often, but I stressed the importance of him reading everything, in order to give feedback, corrections, plus any new thoughts the reading might spur. I wanted him to be especially vigilant for any misinterpretations I might make. He agreed to my methodology, and thought it a good idea to allow George to read each chapter, to give a second opinion.

I asked for three of his earliest memories, explaining I would later analyze them for his attitudes towards women, towards men, and to get a sense of how he saw himself fitting into the world. After analysis, I promised to check with him on the accuracy of the results.

His first memory was of a scene that included his father, mother, and siblings. The year was 1910 and they were all in the fields surrounding their home, celebrating the arrival of a comet—he said Halley's Comet—to the Scottish skies. The name of the comet may have been a misnomer but David's interpretation of the event was still valid for him. What he noted was the excitement generated by the orbiting comet and the feeling of success and pleasure in the family's energetic activity. Later, he learned that some people had been frightened by the comet. "But," he said proudly, "we were not like that."

My training in Carl Roger's "reflective listening" techniques also helped encourage Caldow's thoughts—he answered my questions directly and without hesitation. By the end of our first session, I began to believe our collaboration could work.

David and I continued to meet over the next five years, much longer than I had planned. I fell in love with the story, realizing it was not just David's story but a distinctly Canadian experience, a humorous adventure, and a love story—not only of a man for a woman but also a story of love for life itself. My research helped to reveal David's courage, personal growth, and his contributions to the unique culture that is Canada, helping me realize that others—fellow immigrants, Canadian history buffs, members of Scottish organizations, social history students, anyone who works within a bureaucracy—would appreciate and enjoy his character and the Canadian immigrant experience he described. Thus my sense of audience grew as his story developed.

Large hanging files of information gradually accumulated, often sorted by the "cut-and-paste" method (which was necessary, for example, when details about Scotland were revealed during a segue from an anecdote about Tranquille or Tanzania). When there were "holes" in the progression of a story or if descriptive details were scant, I created lists of questions for the interviews. Sometimes I typed the questions—in capital letters— right into the text I shared with David. His reading and rereading of my writing prodded his memory throughout the process. Thus, specific details did not necessarily arrive in chronological order. Over the five years we collaborated, I cut, pasted, and added details culled from our subsequent conversations, making written descriptions more vivid and detailed whenever possible. I must admit that at times it was difficult to avoid author intrusion. I tried not to allow my ideas to trespass into David's story. It is for the reader to judge how well I succeeded.

A sample of my interview notes and a description of how they were used might be useful to understand how collaboration produced this version of David's life. For example, during my initial interview about Tranquille (chapter 13), David spoke of his and Peggy's arrival, the reasons for and reservations about going. These are my verbatim notes:

> Anyway, when he (Pete Moore) asked me, I said right away, "Sure, I'll go." They trucked me up ahead of time and Pete told me to start right away, just to get to know about things. Peggy was with me. Pete told me, "You'll have an uphill time, Dave. But I know you can take it. There's been trouble there. We've been spending too much money." He introduced me to the doctors and the other people I'd be working with before I started. They had already fired the man whose job I was taking. They gave him two months' pay because he was helping himself too. The people that were left were all very nice people. Of course it was the butcher that was giving it to him. Pete told me about this old butcher. He slaughtered 3 beef and 16 pigs every week. The pigs were butchered and we made hams and bacon and it was supposed to be for the hospital. This old butcher was selling a quarter of beef a week, plus hams and bacon, to a chinaman who had a store in town. Then the butcher was pocketing the money, of course.

In a subsequent interview, I asked why he had decided to leave Colony Farm to go to Tranquille. He replied, "I left because I couldn't stand the new superintendent. I would have quit, but Pete Moore said, "Hang on for a little while Dave. I might have something for you."

Over time, I asked questions about how he and Peggy felt about the move, the new setting, the intrigue surrounding the corruption; I asked him to describe the people involved, the process of applying for government money for the farms, etc. I gave him an article from the *BC Historical News* describing the buildings and the setting. (As well as adding his own details, he also remembered romances flourishing in spite of restrictions.)

After visiting the provincial archives, telephoning libraries, and contacting people in the areas he named, I filed research along with my laptop notes and added this information to the narrative for several reasons. First, David's anecdotes and actions became more understandable when they were embedded in context. For example, when discussing life in Montreal, one day David said disgustedly, "Women weren't supposed to know anything about condoms, yet they were available from drugstores for any man who wanted to buy them to use with his girlfriend!" Since I had recently read Pierre Berton's writing about these years, David's expressed attitude sounded light-years ahead of the society. Therefore, I cited Berton's book to demonstrate this contrast in attitudes and permit the reader to discover a greater truth than David's statement alone would reveal.

I also used research to verify the accuracy of his memories. (For example, David did not know the name of the tranquilizing drug used on patients after they were no longer permitted to work in the fields of Colony Farm, though he knew that the drug "turned their faces blue." This statement seemed bizarre and—to my non-medical ears—smacked of hyperbole. However, in a later conversation with a nurse, I not only discovered the name of the drug [chlorpromazine] but also realized David's observation was correct.) Always, after David had read a new section of the story, I had him respond to included research, asking him if he felt the additions were appropriate. If he did not, I made changes.

I followed a similar procedure with dialogue—a necessary but difficult-to-validate story element. Although David Caldow is a natural storyteller who often includes dialogue in his accounts, my problems with the truthfulness of remembered dialogue are twofold. First, it's impossible to check the recall of exact words when one is remembering over a span of more than eighty years. (For me, truth in dialogue lies in the reproduction of the gist of what was said, as well as the intention and emotion with which it was spoken.) The second problem with dialogue relates to the style of speech, including considerations of grammar and dialect, and how much of it to include. Because I was working with David over a long period of time, I was capable of capturing his manner of speech, but I chose to write only enough to give a

flavour of his conversational style. (For the reader, translating too
much brogue can prove tiresome.) However, the inconsistency in
David's verb forms (e.g., sometimes "we was" and sometimes
"we were") is included because it's true to his style of speech.

Although he is a farmer and did not attend school for long,
David is also an intelligent man who worked for many years, not
only with others like himself but also with many men with doc-
torates. He admired these men and learned from them, so his style
of speech, like his social criticism, became more sophisticated.
Sometimes conversations he reported as narrative (e.g., he told
me that...and I heard...) were written as conversation to create a
more interesting read. But whenever this was done, I asked
David, "Is that pretty much the way it happened? Does that conversation ring true for you? Have I written anything that you or
the other person would not have said?"

For example, David spoke to me of choosing "pay-as-you-go"
for his sexual fulfillment while living in Montreal. Then, in
response to questions, he described his situation, the people, the
laws, and the prevailing attitudes towards prostitution. He said he
chose prostitutes because he "didn't want to make someone pregnant." Later, he gave me his impression of the sex trade in New
Westminster, BC—in 1936, before he was married. These are the
exact words I typed as he spoke to me:

> In New Westminster, there was a house of prostitution on
> Columbia Street from 1936 on, until I got married. It was a
> busy place, and with a bootlegging joint as well. Christ, the
> truck would back up to the door and start unloadin' the beer.
> About every second door on Columbia Street was a boot-
> leggin' joint. The woman who ran the house was well
> known, her name in the paper whenever she was fined.
> Word was, the $300 (the fine) went to pay her taxes. She had
> two daughters in university in Seattle, because she told me.
> We'd sit and have supper and have a good chat. Her name
> was McDonald. There was no rough stuff in her house, it
> was a good house. She was a big, strong woman and if any-
> body'd act up, she'd throw them out herself. It was a hotel.

Clean. The prostitutes made the beds and worked the hotel. [Then he repeated:] I chose this rather than make women pregnant.

David's Montreal tales are found in this story but this New Westminster anecdote is not, simply because (to my mind) by the time it was revealed it didn't fit into the flow of the already written story. Instead, I've filed and marked this one for possible further research.

Had I used it, I would have asked him a number of questions. I would have asked him for remembered details about the house itself, the woman's appearance. I might have included a dialogue with her telling him that her daughters attended university. I would have researched the New Westminster archives for newspaper stories about this place, and perhaps for the laws of the time. Then I would have had him critique the writing, as I did throughout the book.

David proved so forthright that I trusted him to say exactly what he thought of my writing. I made many changes to concur with his opinions as we proceeded. My methods of detailed collaboration take a great deal of time for both informant and writer, but the veracity of the writing and the truth of the man revealed depends upon them.

As David's story grew beyond the expected three thousand words, I was amazed and intrigued with how little I'd known about this man who had—about thirty years ago—given my family its first dog. I'm grateful to him because I learned a great deal about my country and my province from writing this book. Privately, I even became somewhat comforted by the differences between his perspectives and my teenage reactions to the provincial mental hospital.

After he and George had read and approved the words for the last time and selected the pictures, the finished manuscript was formally presented to David at his ninety-fifth birthday party. After the celebration was over, many of the Canadian Caldow clan read the story and commented on it, with approval.

David has reached the age of ninety-eight. He has since moved from his White Rock apartment (by his own choice) into a care home in Surrey. When his son George retired from his position as Personnel Officer for the Coquitlam School District, my husband and I attended the retirement party. Along with many friends, we applauded both George and David as they, and other honoured guests, ascended an impressive curving staircase, following a kilted piper into the large reception room. The senior Caldow enjoyed himself the whole evening, laughing and predicting he would live to be one hundred.

It is for the reader to decide if this Scottish-Canadian saga is worthy of the five years we took to produce it collaboratively. It is more than a farm memoir. It is more than one Scottish-Canadian's story. It is a very human story of the struggle we all—both men and women—experience as we mature and reach for our childhood dreams. It is a story of coming of age in the early twentieth century, a story of love, of the importance of fatherhood. It is a story of achievements in British Columbia's and Canada's history, as well as a story of one immigrant's Canadian experience. David needed me to both organize and write his memoirs into a creative documentary, but the truth of this man needed no embellishments.

To the best of this author's ability, David Caldow's story is true and verified by the man who lived it.

ACKNOWLEDGMENTS

This book came to be with the help of many people. Thank you to the 1994 West Coast Women and Words Conference, and my first writing teacher, Susan Riley of the *Ottawa Citizen*, who gave me the confidence to continue writing. Thanks to my friend Debra Swain, my car-pool-captive audience. Thank you to the members of my Surrey Continuing Ed writing classes and my teacher, Ed Griffin, who read and critiqued many chapters of this book, and thank you also to friends John McKittrick, Jerry Falk, and June Cleghorn who read and critiqued the entire manuscript. Members of my writing group gave me encouragement, especially Dr. Beverley Greenwood, who let me benefit from her expertise with proposal writing. Librarians and archivists from the White Rock Public Library; the Claresholm, Salmon Arm, and Kamloops museums; and staff at the BC Provincial Archives all helped with research. Special effort was made by Janet Finlayson at Macdonald College Library of McGill University, Anita Bedore of the Marmora Historical Society, and Dr. Ernest Peters of Vancouver (a retired expert in extractive metallurgy), all of whom cheerfully shared their time and knowledge with me. Thank you to David Caldow's friend Libby McGinnis for sharing her photos and tales of Africa. Special thanks also to members of the Caldow extended family, especially Geordie, his wife Joan, niece Jean, and husband Gino, who all gave ongoing, caring assistance. Karen Breckenbach of the Colony Farm Park Association not only read and critiqued the story but also reviewed it for the association. Dr. Jack Little, professor of history at Simon Fraser University, read and critiqued the manuscript, helped me find a publisher, and has created an opportunity for me to promote the book. Thank you to Brian Henderson, Leslie Macredie, Penelope Grows, and my painstaking editor Jacqueline Larson, of Wilfrid Laurier University Press, all of whom welcomed me and made me believe this book would become a reality. Thank you most of all to David Caldow who bravely risked sharing his very personal Scottish-Canadian adventure. To all these generous, positive people and finally to my family, especially my husband Tom, my gratitude for your patience and knowledgeable support.

PROLOGUE

The early evening sky is divided by a trail of light covering two thirds of the distance from the horizon to directly overhead. The highest end of the light trail shines as brightly as the brightest of the twinkling stars surrounding it. Silent, unmoving, an awe-inspiring splendour.[1] I'm with my brothers and my sisters outside my home in Scotland, looking up in wonderment at Halley's Comet, visitor to the skies of Scotland in 1910.

Even more impressive to me is the sight of my father loading, tamping, and firing his muzzle-loaded rifle, over and over again, a traditional Scottish celebration, this time for the comet ablaze overhead. He is a powerful man. Each time he loads, he leaves the tamping rod in the barrel. When he fires, the "whoomps," of the fiery blasts send the steel rod arcing, careening into the twilight of the fields surrounding us. I see mother smiling, watching me charge through the grasses with my brothers and sisters, searching for the rod. We call to each other, our calls filled with laughter. When we find the rod, we run effortlessly, returning it to father. This is my earliest memory and, though I'm now in my ninety-ninth year, I can still feel the joy of our success.

I then had two dreams that seemed at odds: travelling like the comet to alien places, and becoming a farm manager like my father, attached to the land. In a time before passenger flight or freeways, achieving both ambitions seemed impossible, but the chasing of these dreams became the story of my life.

Note to prologue is on page 239.

CHAPTER ONE

Scotland

. . . O Life! how pleasant in thy morning,
Young Fancy's rays the hills adorning!
—Robert Burns, "To James Smith"

My name is David Forteath Caldow. My family lived in Dundrennan, a little village near the big town of Kirkcudbright (we pronounced it "Kircoobrie"), the capital of the shire. Kirkcudbright wasn't really a big town, but we thought it was big. We were proud of the fact it was known as the place where Robbie Burns died. The fact that someone died there may seem an odd motive for town pride, but if you're Scottish, anything to do with Robbie Burns makes you proud.

I was born at home in Gatehouse, Scotland on August 15, 1903, one of the middle children. Of the five of us, I was the only one fed by bottle, cared for mostly by my sister, Mary. My mother couldn't nurse me because, I'm told, she had "beelin' breasts," what we'd now call "infection." Later, my father Joe Caldow, my mother Jane Maltman Caldow, my five brothers and sisters, and I shared a three-bedroom house, located about three miles outside of Dundrennan. With only three bedrooms, you'd expect the house was cramped, but we were never all of us in it at the same time. You see, as each of us reached fourteen years old, we were automatically out of school and off to work. Often we boarded where we worked, which was usual in those days.

Our wee house was surrounded by a big vegetable garden that my father had tilled using plough and horses. As well as vegetables, we grew red and green gooseberries, lots of rhubarb, and big fruit trees. I remember sometimes our apples were spotted, but we ate them gratefully anyway. If any of us children were caught picking and eating without permission, we got a good going over because all produce not eaten fresh was slotted for jams or preserves.

Those of us small enough to live at home always had a job to do when we weren't in school, even on Saturdays. It was then we went into the surrounding countryside picking wild gooseberries, raspberries, and sloes. When the season wasn't right for picking, there were other jobs found. We gathered the eggs of birds we called peeseweeps, or lapwings, selling them for maybe a silver shilling apiece. The eggs were valued for their use in making medicine. Sometimes, we'd walk the fields finding wool, shed naturally from the sheep. Mother took this wool to the local wool mill, to be made into rough cloth for our trousers and jackets. We all worked hard and were proud of our achievements.

Mother took us to church every Sunday morning, then to Bible class at night. I didn't understand much at church, especially when the preacher ranted about the glories of heaven and the fires of hell. Comments and spankings had already convinced me I was bad, so for a while the preacher's sermons made it seem certain I'd be going to hell.

Scottish storms seemed a form of this hell to me. I was terrified of the roar of thunder when I was small, even though mother and father told me not to be afraid, there was nothing to it. We little ones told each other it was God making thunder because He wasn't pleased with what we were doing, and we believed our own stories.

When I grew a little older and observed the "goings on" of some of those considered pious, I'd say to myself, Well, if they're going to heaven, so am I. Then I began to feel better about myself. (Nowadays, I don't go to church. I believe people have to be good to one another, but I don't like all the "politics" that comes with churches.)

When I was about ten, I had a job walking the big stallions for "The Banks," a farm owned by Drew Montgomery, one of the most famous breeders of Clydesdale horses in Scotland. His Clydesdales were selectively bred. The desired characteristics are similar to those modelled by a Shetland pony: nice flat bones, hair on the legs, an ideal body shape, but much larger. Their size gives them the strength and stamina needed for heavy work. To anyone's eyes, Clydesdales are very large horses, but to me, at ten...my! They were huge!

By the time I was sixteen and working at Castle Creavie, near the town of Castle Douglas, I had learned to break riding horses. Mind, these were farm horses, used to being handled by people—just never ridden. They were nae hard to break. I learned early, in breaking a horse you must use kindness to get them to obey. People can create terrible problems using physical punishment to break a horse.

Before you climb on a horse's back to ride it, it's important to introduce the use of a bit for guidance. A bit is the metal part of the halter, placed over a horse's tongue to put pressure on it when the rider pulls on the reins to give the horse direction. A mouther is a training bit, made rough so it feels uncomfortable on the horse's tongue. A mouther is used before a regular bit, but only for a short time. The brief discomfort caused by the mouther is useful to the training. After the exposure to the mouther's roughness, a regular bit then feels comfortable to the horse and will be accepted. One horse at Castle Creavie had been hurt, because a couple of men used a mouther the wrong way.

I had to nurse this poor horse back to health before I could begin to break him. The two fools had hitched the horse to a wagon out in the field, using a mouther instead of a regular bit. When pulling the wagon, the horse damn near had his tongue cut in two by that mouther. Then he had bolted with the pain and ran over some logs, injuring his legs. The poor horse was so spooked nobody could do anything with him. If I'd called the police on those fellas, they'd have been thrown in jail.

The first thing I did, I washed the horse all over. I did this every day until his legs healed. With the comfort I brought him, he began to trust me, and eventually he let me ride him. When I finally hitched him to a two-wheel cart, I had no trouble at all. He became a good horse, but it took kindness to make him that way.

It was about this age, I began to appreciate how hard my mother worked, spending very long hours to provide not only for us, but also for the men who worked with us. We never ate bread, it was always the scones she made with rich, real buttermilk—not like the stuff that you get in stores today. At times she had to bake scones twice a day, as well as do all the housework and cooking.

Mother was also a midwife who helped the dairyman's wife birth their eleven children. When the dairyman's wife died after the birth of her eleventh child, her eleven-year-old daughter had to take her mother's place, rearing the other ten children. My mother went over to their place often, to check on the girl and help her out. Even though mother had a heavy load of work herself, she would always find time for someone in need.

I began to appreciate my father's efforts, too. It was from him that my younger brother John and me absorbed the love of farming which became our lifelong occupation. From father we learned the ways of the animals, the beauty and the danger in being close to nature. For the rest of our lives, we used the skills and the values he taught us, never really feeling fulfilled working at anything except farming.

John, who was only five when I left home, would have had a very short work resumé, if he'd ever the need to produce one. Leaving school at age fourteen, he worked on only one farm, for three generations of the owner's family. At eighty years old, after suffering a heart attack—caused by him running around to get the owners' tractors out of their burning barn—he retired from working for the grandson of the man who first hired him.

Love of farming was not as enduring for all my brothers. My oldest brother Jimmy joined the Scottish police force after World War I. Bob first worked the farm in Scotland, later joined the police force, then left Scotland for Canada where he became a successful fruit rancher. Will became a carpenter. All of us were relatively successful at the work we chose to do.

My father, in his day, was known as the best farm manager around. He became a manager not only because of his skills, but also because he worked harder than any other hired man. People knew he had earned the respect given him and other men wanted to work for him, even though he was a hard taskmaster. He never lacked for a job because owners of all the big estates around Dundrennan knew his reputation.

In spite of his success, my father never owned a farm himself. The farmland in our part of Scotland was all within the inherited estates of the feudal system. Unless you were born into landed gentry, you farmed for someone else.

My uncle Jake Mulraith, my auntie, and their three boys, Jim, John, and Robert, farmed sheep for the owner of a three-thousand-acre place in hilly New Galloway. Jake and his family lived in an ancient stone farmhouse that sat in a row with a stone barn and chicken coop, the buildings surrounded by hillside meadows filled with hay. Three sheep dogs helped Jake look after from three hundred to five hundred sheep, depending if the lambs were grown. At shearing time, my father and the rest of us, along with other sheepmen in the area, would go to help Jake. In return, at lambing time, Jake would bring his sheepdogs and help us. Jake was a man who never did anything else but look after sheep, all his life.

From watching him and the other good dogmen in Scotland, I learned to love training and working sheepdogs. Eventually, my father gave me my own dog, "Sweep," a good dog. I learned to prefer working with bitches because, once trained, they could be relied upon. I don't think you can say the same for the human variety. My interest in sheepdogs and their training lasted for the rest of my life and led me to interesting challenges later on, in Canada.

Shearing time in Scotland was a sight to behold. Seven or eight sheepmen worked together, using their dogs to herd thousands of sheep into shearing pens. Once in the pens, the sheep wouldn't be fed, due to the problem of collecting, storing, and distributing a large enough quantity of grass. The challenge for the sheepmen was to get those thousands of sheep shorn quickly, then returned to pasture before they starved, within three or four days at the most. The only way to meet this challenge was to help each other, all of us working as fast as possible.

Among these men, my uncle Jake was one of the best, because they'd timed him once shearing five sheep in seven minutes—with handclippers! He could shear holding clippers in two hands at once. Sometimes others would stand watching him, amazed, cheering him on as his skill kept the fleece flowing smoothly away from the skin of the docile sheep. Our whole family was proud of Jake's prowess; even my father spoke of him with admiration.

When we worked together, my father was to be heeded with no talking back, that's for sure. Obedience was a lesson I learned well, because I worked for him longer than anyone else in the family. During the First World War, he took me out of school at thirteen, to replace three hired hands conscripted into the army. He insisted the two of us do the same amount of work that the four of them had done.

When plowing, he taught me how to control the horses, how to hold the plow just right to keep the soil packed so the seeds fell in the best way. He told me always to be on the good side of the horses. I had to treat them right, in order to have them help me meet Father's expectation that we finish in half a day, the same amount of work he had previously expected a grown man to finish in a full day. Ploughing for my father was heavy, hard work. Mother realized he was working me so hard he was nearly killing me, so when I turned fourteen, she made him let me go to work for our neighbour, Mr. John Cruikshank, farm manager at Castle Creavie.

Castle Creavie wasn't really a castle. It was a farm on the site where an old castle once stood. All that was left of the old castle was a large mound and a moat. There was also the remains of an old Roman road nearby. At Castle Creavie, I exercised, groomed, and trained Clydesdale show horses. When I took this job, I had no idea it would lead to later good fortune in Canada. I also didn't know the Cruikshanks would eventually become my in-laws, when my sister Jess married into this family. At the time I was just happy to know Castle Creavie offered me an easier job than working with Father.

Mother arranged for me to be paid six pounds every six months, plus room and board. At that time, grown hired men made one pound per week—four times my pay. However, I didn't mind the low money. Any job for any pay was better than going to school. I didn't miss being a student at all. For me, school had been nothing but trouble.

> We. J & B Cruikshank have had David Caldow in our employment since leaving school about 7½ years, and have found him to be a good - plough human and able to work all machinery used on a farm, a good milker and able to do anything about a farm and have always found him honest trustworthy and very obliging
>
> J & B Cruikshank
> Castle Creamie
> By Castle Douglas
> Scotland

The year was 1922.

The first trouble was the walking. We had to walk about four and a half miles to get there, spend the day, then walk back four and a half miles. All this walking sapped our energy, but this didn't excuse us from nightly chores. Three of us, Joe, Bob, and me, all walked to school together.

Once at school, an unsmiling, miserable headmaster enjoyed hammering us, often for reasons that made no sense. For example, he once strapped me for lying, not believing the true reason I gave for my brother's absence. Joe had cut his foot on glass, easy to do because we always went barefoot in the warm weather. My teacher, Mrs. Murray, gave the headmaster an earful after I returned to class and she found out how I'd been treated.

Another time when we were particularly fed up, my brothers and I tried skipping school for three days. On the third day—when we returned home pretending we had been at school—my mother presented us each with a scone, saying, "You'd better eat this because when your father finds out you've been skipping, you'll get no supper tonight." We had no idea she'd had a visit from the "whipper in" (the truant officer), earlier that day.

That night my father caught Joe first and hammered the hell out of him with a big horse strap. Joe was tough, I'll tell you. He never cried, no matter how much he was hit. Then my father caught Bob and he got it, but not as bad as Joe. Maybe because Father was worn out himself, he stopped when he got to me. But by this time I was out of it with fear, anyway. Later, in our bedroom, when Bob and I were crying hard, Joe looked at me and said, "What are you crying about? He didn't hit you!" Needless to say, Joe was bruised and shaken and it took him several days to get over the beating.

Now, I want you to know, I'm not blaming my father for the beatings he gave us. We were bad, I'll tell you. We deserved everything we got and we got it often. If we were hammered at school, we'd get hammered at home for getting hammered at school. So, perhaps you can understand why, to my mind, school was nothing but trouble.

Other times, my parents were kinder. I never once heard an angry word pass between my father and mother. She never opposed him when he disciplined us. On Sundays, the only days he didn't work, he rested. Only rarely did he play with us. He never lifted a finger around the house, never lit a fire or boiled a kettle. Dad budgeted money for the farms he managed, but Mother looked after the household money. When we were paid on other farms for our work, we brought our money home to her. Mother bought all our clothes, having everything tailor-made, even our shoes. When we were older and working away, we came home on Sundays for tea and scones with our parents. I remember having long conversations, all of us together, mostly about farming and the people we knew. Those were the good times.

When the First World War ended, both Mother and Father wanted me to return to the classroom, but I refused. I was seventeen by this time, and I couldn't see myself in school, sitting with all the little children. Because I'd been doing farm work for some time, I thought I had become a man.

By this time I was prowling with the lasses, just as any seventeen-year-old Scottish lad would do. Everybody knew there was ways of finding pleasure when you needed it, and needing it was only natural. With animals all around you on a farm, you couldn't live and not know that. Just the same—maybe because mother was a midwife—I had my principles. I was careful not to make anybody pregnant, and I didn't take advantage of girls who weren't willing, either.

Lots of fellas used a girl on a nearby farm. Hell, it was common knowledge, you could take your pleasure with her any time you liked because she was sick. She'd been born a hermaphrodite, she took fits and she was kind of simple, but I always felt sorry for her. I never used her, though many did. I went with lasses who wanted the same pleasure I did. I was beginning to believe I was a man of the world. There was a certain risk to my rovin'—if I'd ever been caught, I'd have been beaten within an inch of my life!

I didn't realize how little I knew, how little I had experienced, knowing only the Scottish way of life in our village. Other cultures, other values, or ways of life were practically unheard of. My only friends without a Scottish brogue were non-human: the cattle, sheep, and sheepdogs, and even they were Scottish kin. There was no television to bring the world into our cottages. I never saw a streetcar until I was twenty years old, when I went to the city of Carlyle near the English border. The first time I rode on a train, I thought the train was out of control, running away with me. It seems funny now, but not then. Yet, I dreamed of taking off on that train and circling the globe like Halley's Comet. My sister Mary provided me the unlikely opportunity to begin.

Much to my parents' dismay, Mary had fallen in love and married our cousin Tom Twiname, whom my family had practically raised. After living with us, he had emigrated to Canada, lived there for three or four years, joined the Canadian army, then returned to Scotland as a soldier during the First World War. In spite of our parents' objections, he married Mary, then reluctantly left her behind in Scotland, when he was ordered to return to Canada with his army unit.

Mary asked me to accompany her and their wee son Oliver on their journey to rejoin her husband and begin life anew in Canada. I couldn't wait to go. My mother agreed it was a great opportunity for me, but my father said nothing until the day I left.

The words he spoke on that day became the centre of my thinking for the rest of my life. My father said, "Dave, here in Scotland you've always did what I told you, but in this new country the people might not tell you what to do. They might expect you to know what to do without being told. When that happens, you think of me and do the job the way I told you—and mind you do it right!"

I knew what he meant and his words gave me confidence. Because I'd worked for him and shared his wisdom, I knew his teachings would give good guidance any time. I believed I could work for anybody anywhere and do well. I had no idea of the challenges that were to come, once I left the certainty of Dundrennan. Nevertheless, with my letter of reference from Castle Creavie in hand, I boarded that train.

CHAPTER TWO

Beginning the Trail

Wi' quaffing, and laughing,
They ranted an' they sang;
Wi' jumping, an' thumping,
The vera girdle rang.

—Robert Burns,
"The Jolly Beggars, Recitativo"

Mary, Oliver, and I left the train; we made a brief visit with my aunt, Mary Mouran (nee Maltman) and her family. They lived near Liverpool, England, a huge port city, and it was here I began to suspect my youthful confidence might be premature.

Like Liverpool, my aunt impressed on first sight. She was unusually tall, over six feet, strong and confident. As a special treat, her family took us for dinner to one of the fanciest restaurants. I put on my best white shirt with stiff collar and tie, polished my shoes as best I could, yet I still felt out of place. As I dressed, my sister cautioned me.

"We'd best mind our manners here, Dave," she said. "These are city people."

In the restaurant, I sat down at the table and, Jesus! There was all this cutlery! I felt embarrassed and panicked a bit, but did my best not to show it. Aunt Mary must have seen through me, though. She winked kindly at me, whispering, "Just watch me, and do what I do. It will be all right," and so it was. I was grateful and I've always liked her, ever since that meal.

After the dinner, my thoughts circled back to my father, feeling gratitude for his help. I made up my mind to buy him a gift next morning, a decision that could have landed me in an English jail. Except for Aunt Mary's intervention, I think that's where I'd have been.

The incident happened after we entered a Liverpool tobacco store, where I hoped to purchase a pipe as the gift for my dad. Standing in the scent of imported tobaccos, looking down the rows of pipes in the glass display case, I spotted a beauty I knew my Dad would take pride in puffing. It was a curved and polished dark brown walnut Meerschaum, with a little rim of brass trim halfway down the bowl. I pointed to it and handed the English salesman my hard-earned money.

"I'll take this one," I said.

He stepped back from my offered hand as if it held something dirty. Looking down his nose at me, he said gruffly, "Scottish money's no good here, lad. We only take English treasury money." (He emphasized the words "English treasury money," as if he were talking to a simpleton.)

When I heard his words I saw red. I felt the heat as my face flushed, my fist clenched and my feet braced themselves for a fight.

Within the moment, I also felt my aunt's calm hand touching my forearm, nudging me gently towards the door. With her quiet urging, we stepped quickly outside, where she lead me to a nearby bank. With no fuss whatsoever, she changed my money, then went back and bought me the pipe. She was quite a woman, my Aunt Mary Mouran.

Next day, with my aunt and her family waving to us, Mary, her two-year-old son Oliver, and I boarded the CPR ship S.S. Montclare, one of three sister ships. It was beautiful, big, gleaming white, and to our eyes, very modern. As we climbed the gangplank, I was excited about boarding, but once on board disappointment set in. Instead of leaving for the rest of the world, we were unexpectedly returning to Scotland, again. Backtracking was necessary to pick up passengers who had arranged to board in Scotland, not Liverpool as we did.

When we reached the Scottish pickup point, Mary and I stood at the Montclare's rail, watching the arrival. We were surprised to see the newcomers loaded by barge, bobbing in sunlight reflecting on the waves, looking excited and noisy as seagulls riding on driftwood—except, these seagulls had colourful carpet bags and leather luggage piled around them. From the crew, we learned the barge ride was necessary because a longshoremen's strike prohibited ships from docking, limiting ship access all the way up the Scottish coast. We were told no cargo was permitted to be loaded, which was only an interesting fact, not worrisome to us at the time. There was no way we could know that our lack of cargo would have consequences for us later.

For the first couple of days, I stood often at the ship's rail sniffing the salt air, feeling excited about experiencing life on the ocean aboard this huge ship. However, by the third day my enthusiasm evaporated. The sea changed from calm to stormy and the ship heaved on the massive waves that assaulted her day and night for the rest of the trip. The emptiness of the cargo hold left the ship free to move and roll as she pleased, being low on ballast weight.

Did that rolling create misery! From the third day aboard until we docked in Canada, the ship's rails were lined by seasick people, clutching them for support as they vomited their meals into the wind and spray. Not that I was seasick myself. I was luckier than most. I was twenty-one years old, strong from all my years of heavy farm labour, and full of energy. It was my sister Mary who took to her bunk, providing me with my very first experience of babysitting.

Let me tell you, caring for that wee Oliver gave me the surprise of my life. Jesus! He needed not just one, but three energetic men to look after him, he was so active—and he only had me. When Oliver got himself lost, I felt only disgusted, not guilty. But you can imagine, my sister was greatly upset.

"What?" she almost screamed! "Ye've lost him?" as she climbed out of her sickbed. "Oh, my wee Oliver! Don't stand there gawkin' at me, ye fool. Get back out there and find him!"

With her feeling frantic and me very annoyed, we hunted up, down, and everywhere on the Montclare, but he was nowhere to be found. I can still see Mary as I came upon her out on the deck, staring out to sea looking so sad, thinking Oliver was drowned.

I told her straight out, "Mary, he's a lad that's much too bad to be overboard." (*Only the good die young*, I thought.) "There's no use fretting yourself."

And I was right. We finally found him happily hopping and scrambling over chairs in the dining room, having no idea at all what we'd been through because of him. He was so bad he made me question whether I'd ever want a son, I'll tell you, and his poor mother seasick the whole time.

Knowing the sickness almost everyone was experiencing up on our deck, I was glad our cabins were up top. There were hundreds of "Europeans" (not British subjects) quartered by regulation on the lowest deck. They ate in their own lower dining room and were never allowed to mix with us. The regulation achieved its purpose, making them almost invisible to us who were considered British subjects. They were on this ship to escape the aftermath of the First World War in Europe. Much like us, they were setting themselves on this journey to build new futures in Canada. Sometimes I thought about their misery, having to endure the hell of being tossed in the belly of the Montclare for over a week, but mostly I thought about how glad I was not to be European.

My first really close encounter with what I considered aliens, occurred in my cabin. I bunked with two Welsh men (down-to-earth miners), an Englishman and a Cockney, who spoke in rhyming slang. Because we each used different versions of English, only the two Welshmen could really understand each other. Living with these characters seemed a little like living in a circus. They were the animals that entertained me, and I suppose they had similar thoughts about me. We spent a good deal of time laughing hard and long at our fumbles and our misunderstandings. I began to appreciate that the sharing of differences could be fun.

I was seated at mealtimes with another group of four fellows, about my age. We all loved whiskey, and all felt discomfort with the sophisticated atmosphere of the ship's formal dining room. We all did our best to change this formal mood.

A Scottish lady—my sister's cabin companion—helped us all to relax by giving me a bottle of Dewar's Scotch. (She knew it was illegal to take whiskey into Canada because prohibition laws were in force, and our ship was soon to be easing into Halifax harbour.) I shared her gift with my four dining companions. We used it to celebrate the end of our sea journey, downing the whole bottle between us, at lunch. As we tipped our glasses to lick the last drops, all five of us were crazy with booze and with excitement about our arrival. Our farewell to the Montclare included gleefully throwing handfuls of slushy mashed potatoes at the snooty, shocked, dining room stewards.

Escorted forcefully through the dining room exit, we understood their message clearly, in universal language: The staff was not amused. So much for our sophistication, but we didn't care! We'd had our fun. Now we were off to the land of promise: Canada!

Mary, Oliver, and I walked down the gangplank with a mixture of feelings—excitement, curiosity, and some concern about things working out. Mary set me the job of buying our box lunch, which had to last us all the way to Montreal. We learned there'd be no dining car or bunks for us on the train. The three of us would be sitting up all the way.

I had to push my way through the crowd until I got to the lunch counter. While I was ordering, a group of the Europeans from the ship surrounded me, pushing me, yammering loudly at the woman who was serving me. She stuck her fingers in her ears to block the noise. Garlic wafted my way from them, the first time I had ever smelled it.

Jesus! I thought, *I can't believe people can smell so bad.*

I bought our lunch and got out of there as quickly as I could. Not knowing any better, I went back to my sister and said, "These Europeans is all rotten! They stink!"

Later in my working life, I learned I'd been wrong to judge in ignorance. These Europeans were probably good people, and much needed in Canada (though nothing has changed my opinion of garlic. It was a long time before I tasted it, and I still don't like it very much).

When we arrived at the Montreal train station, I was remembering all these stories I'd heard back home about robbers, Indians, and the like in Canada. About this time, the redcaps innocently came over trying to help me with our bags. (And mind, I'd never seen a Black man until these redcaps came my way.) As they were doing to everyone, they picked up our bags.

With a lunge, I grabbed the bags back, yelling "Stop!" louder than any European in the station. I thought the redcaps were thieves trying to make off with our belongings. The surprised look on their faces and their laughter made me realize my mistake.

My sister Mary's husband Tom was watching me through the gate and laughing like hell! The redcaps must have thought I was crazy and I felt very foolish. I hoped my future in Canada held more than embarrassment, but I needn't have worried.

CHAPTER THREE

Quebec

As he was walking up the street,
The city for to view,
O there he spied a bonie lass
The window looking thro'.
—Robert Burns,
"Charlie He's My Darling"

Initially, in 1924 I lived with Mary, Tom, and Oliver in their Montreal duplex. My brother-in-law continued to be a devil, always finding ways to enjoy himself at my expense, just as he'd done at the train station. For example, he took me to dinner in a Chinese restaurant, then guffawed at my reaction when I saw my first Chinamen. When we walked into the restaurant, I was taken aback, I'll tell you.

"Christ, there's no place in Dundrennan like this," I muttered.

Because I thought the waiters looked so strange, I didn't trust them, or their food. Tom chuckled all through the meal because— even though I was hungry—I couldn't bring myself to eat much. The reaction was pure ignorance on my part, of course, and I deserved his teasing.

Years later, after working with Chinese at Kelowna, some became my good friends. For the most part, I came to know they were kindly, honest, hard workers. I learned to like the people, but I've never learned to like their food, especially after they started making the spareribs red.

For almost four years, I lived in the village of Ste-Anne-de-Bellevue, a town close to Montreal, located on railway lines leading to Toronto and Ottawa. More important to me, it was close to my cousin Bessie (sister to my cousins in Secum, England) and her husband, Bob Johnston. He owned a store in Verdun, where I sometimes visited and helped out on weekends.

Note to chapter 3 is on page 239.

I was lucky to get the job working on the experimental research farm of Macdonald College at McGill University. Here, entering students who had proven they knew English, elementary mathematics, history, and geography, took training to qualify either to teach school, or to continue and earn the degree of bachelor of science in agriculture.[1]

I worked for the agronomy department where they were interested in experimenting to find the best dates for planting, rates of growth, methods of seeding and crop breeding: developing new strains of cereals, hay and pasture crops, plus root and corn crops. I also worked for the Department of Animal Husbandry, where sheep and beef were raised, not only to feed the students but also to study animal nutrition, diseases, and breeding. A hundred head of milk cows were maintained, with some of the milk consumed by staff and students of the college and some of it sold and delivered to townsfolk in the village.

In Quebec, the only time I ran into trouble because I couldn't speak French was when I was given the job of delivering milk to one of our good customers, the wife of a store manager. No matter how hard I tried, I couldn't understand what she wanted and eventually I had to call her husband for clarification. She wanted four quarts of milk a day and a pint of cream, her usual order. I never did learn to speak French, but in those days that didn't prevent me from being hired, or bar me from progressing on the job.

No matter what kind of English a person spoke, in time you could make yourself understood. Even my cockney friend "old man Ramsey" made out okay. (His name was given him because Ramsey was too old to be working, but he'd lied about his age. He had no pension.) In spite of the language difficulties, he never lost his sense of humour. He was used to people misunderstanding him because of his cockney accent. One day he told us a joke about going to school in England.

His teacher had drawn a deer on the blackboard, but none of the cockney children could tell her the word to name the animal she had drawn. Eventually she gave her students a clue by asking, "What does your dad call your mother when he's talking to her nice?"

"Oh teacher," said my friend, "You can't fool us. That not a picture of a bleedin' old cow!" We all laughed, more because of the way Ramsey told this joke than because of the joke itself.

I had an accent, too, but my accent is Scottish, one of the best accents to have here because, as the name implies, Macdonald College of McGill University was very Scottish. It was founded and mainly financed by Sir William Christopher Macdonald, former chancellor of McGill University, who died seven years before I arrived in Ste-Anne de Bellevue. He must have been a fine man. He had no children of his own, but he put much of his money into improving the schooling for the children of rural communities. He made his money in Canada by starting the Macdonald Tobacco Company, yet he didn't smoke. And like most of us of Scottish descent, he revered the Scottish ways.

As a result, some Macdonald College buildings had Scottish names, such as a main building called "Glenaladale." A herd of Scottish Ayrshire cattle shared the farm with other dairy breeds. Among the sheep, there were South Country Cheviots, originally imported from Scotland. On the farm attached to the college, some Scottish-style stone buildings housed the animals.

At first, because I was single and the priority was to hire married men, the college offered me only a summer job, the least responsible one on the farm. I turned the crank on the machine that ran the mechanized clippers for the sheep. Starting at the bottom didn't bother me, though. In this new land with its strange mix of people, I was glad to work in a place that had so many reminders of home.

In part, Macdonald College hired me because Jack Houston was the respected farm superintendent. Twenty years before me, he had also worked with John Cruikshank at Castle Creavie, in Scotland. Because he understood the requirements for employment at the Castle, he knew I must be a hard worker and well-trained by my father, so he put in a word for me.

Jack Houston wasn't the only man respected by me and others at the College. There was Dr. Crampton, who was kind of a dietitian for the cattle ... experimenting to find the best way to feed them. He also supervised the pigs. He had been a cabinetmaker and had to work to earn his education. Perhaps that's why we became friendly enough that I met his wife. Mrs. Crampton, with the doctor's blessing, offered to help me. She encouraged me to become a student at Macdonald college and complete my education. When I realized I'd have to complete my high school first, I wasn't interested. I only worked with Dr. Crampton when I finished my jobs in the piggery, and still think of him as a fine man with solid knowledge to share.

Dr. Barton, the dean of the Faculty of Agriculture, was another champion person, a real gentleman and a great teacher. Fourth-year students told me he could explain things so clearly they got more learning in fifteen minutes with Barton than they did in a whole month from the rest. I also respected Dr. Barton's knowledge and love of horses, the way he cared for them.

Dr. Conklin was the veterinarian and an excellent animal doctor. He proved himself to me when I took him a urine sample from an ailing horse that shouldn't have been ailing. It was one of a recently purchased pair I'd been given to train, and the farm wouldn't have knowingly bought a sick horse. When this horse first arrived, his skin was so shiny you could have combed your hair in its reflection. I'd seen horses looking like this when I worked at "The Banks," in Scotland. To me, this horse's appearance meant he'd been given arsenic, doped so he'd shine and show good at the sale. Dr. Conklin recognized his condition right away.

But there was one fellow walked into the barn one day, I didn't like him at all. He told me his name was Hamilton, maybe hired as a short-term lecturer to instruct students about raising sheep. I watched as he began to work with them, eventually focusing on a ewe that was giving birth.

I thought to myself, *He's one who hasn't been raised on a farm.* He doesn't have that common sense acquired over a lifetime of living with nature and animals. As I continued to work and watch him, I became horrified. I thought, *I can't believe it! I'm watching him take the afterbirth away from that ewe!* I couldn't stand it. He was doing something I'd known was wrong ever since I was a boy.

I said, "What are you doing? You don't take the afterbirth away from a sheep! You'll kill it. It's okay for a cow or a horse but not for a sheep!"

Maybe thinking he knew better than a farm labourer, he ignored me and continued. The sheep died, of course, and Hamilton learned at the sheep's expense. I wondered how he ever got his job, with the little he knew.

Another time, Hamilton thought he was doing something important enough to start timing himself, as I was turning the crank running the clippers for him, shearing. He was so puffed up about himself, proudly asking me, "Fourteen minutes, now what do you think of that? What would they say about that in Scotland?"

I remembered my Uncle Jake shearing five sheep in seven minutes with handclippers.

Hamilton's face reddened and he sputtered when I replied, "I don't know what they'd say about it, but I know what they'd do. They'd throw you off the farm for incompetence!"

"What do you mean?" he puffed. "What do you mean?"

He looked a little sheepish himself, when I explained about Uncle Jake.

My favourite animal on the Macdonald College farm was a mare named Maggie. I was first introduced to her when I was looking to select a team of horses for the next job Houston gave me, chain-harrowing the meadows. Maggie was skittish, jumping and pulling around like a young mare, but in fact, she was seventeen years old. She had long ago been ruined for work because some fool pulled too hard on the bit in her very tender mouth, bungling her training.

By patting, coaxing, and befriending her, I got her to do some work. Houston couldn't believe it because, up until then, she hadn't pulled a thing. Eventually, she worked a roller for me, on all the fields. I was sad when Maggie was sold to a French farmer who was a mean cuss who hammered her like hell whenever she balked.

Next, I was asked to fix a botched job of seeding oats. This was a problem created by having a man run the seeder who hadn't the advantage of having a farmer father to show him how to seed properly, then insist he do it right (as my dad did for me). The fellow left gaps in the oats, not only wasting growing space, but also creating a field that looked as if it was planted by an amateur.

When Dean Barton saw his work, he said, "Get rid of him. These are experimental fields, used to demonstrate our knowledge of plant breeding. We can't have people looking at this!"

That's when Houston brought me in. After the oats, he asked me to plant corn using a John Deere corn planter that spaced the seeds three feet apart. Because Houston couldn't run the risk of another sloppy job, he made me guarantee I would plant the rows straight and evenly spaced. The promising made me tense. After I finished, I worried about something going wrong, worried that I'd meet the same fate as the other fellow and lose my job. I didn't sleep for the three days it took the plants to come up.

When those little green shoots pushed above the dark ground, Houston came with me to see them. Straight-faced and with no explanation, he turned to me and told me to move my belongings, directing me to a place at the front of a specific barn. Puzzled and apprehensive, I didn't have the courage to ask his reason.

I found out what it was the next day when a smiling Houston announced, "You've been promoted to head teamster. Now you've no worries about keeping your job."

Now, not only Houston wanted me, the dean wanted me also. They gave me a contract saying I had to give a month's notice if I were to leave, or else they could claim a month's pay. In the contract, I also had to guarantee that I wouldn't strike. Now that my job was secure, I began to relax and enjoy the social life in Ste-Anne-de-Bellevue. As well as my sister's family, new friends and new interests helped to make life interesting.

Very soon, my brother-in-law took me with him to hockey games, also introducing me to our neighbour, the father of Pete Lepin, a star hockey player with the Montreal Canadiens. Montreal had two teams, the Canadiens and the Maroons, but I

was never fond of either one of them. My favourite team was the
Toronto Maple Leafs. The reason? I liked the attitude of the
owner Connie Smythe, because he insisted the Leafs never let up
on hard work. I learned he was a racehorse owner too, which
meant, to me, he was a man of good taste. I learned to enjoy
watching hockey games as much as Tom and Mr. LePin did.

My enjoyment motivated me to try playing hockey, skating
with one of the interdepartmental teams at Macdonald College.
Before beginning, I practised a bit, skating on the river. The night
of my first game was very cold and the moon was out, reflecting
off the ice and snow around the outdoor rink as I arrived. A
crowd of spectators, friends, and fans of the Macdonald College
team and their opponents, a French team, laughed and cheered as
the warm-up skate started.

Confident as hell—perhaps because wearing metal-cleated
clogs on the ice in Scotland prepared me for skating—I climbed
over the piled snow and the boards surrounding the rink. Pushing
off on my new skates, I skated hard, getting faster as I went.
Almost immediately, I knew I was in trouble. Those boards at the
end of the rink were coming up too fast. *Too late*, I thought, *I
don't know how to turn!*

I flew through the air, right up and over the rink boards and
piled snow, landing damn hard and flat on my back, hidden from
the other players. Some spectators stared at me, bewildered. I lay
dazed for a moment.

Then I heard my teammates, "Where's Caldow? Where's
Caldow?" they said. "Where the hell did he go?" In spite of the
bright moonlight, I'd been moving so fast my friends hadn't seen
what happened. Though I was shaken, I laughed at myself,
scrambled to my feet, and climbed back on the rink.

Later in the game, I discovered amateur hockey could be
rougher than professional. The French goalie started a fight,
slashing one of our players. I was close, so I grabbed the goalie's
stick from him and began flailing him with it. In no time, all the
players squared off punching and wrestling each other. Pretty
soon all the spectators were out on the ice. Everybody was

yelling in both English and French, but their punches spoke a universal language. What a free-for-all. As we changed to go home, I learned these brawls happened frequently and I thought, *This hockey's a damn good game!*

About a year after I'd come to Quebec, a letter was delivered and, before I opened it, I noticed the handwriting on the envelope was my mother's. Though we kept in touch, she wasn't one to write often, so I wondered what had prompted this note. I unfolded the page and read her few words telling me my father had died, from what my mother thought was pneumonia. That he was dead saddened me, but the cause of his death didn't surprise me or strike me as unusual. (People didn't live as long in those days, so we weren't shocked by death.) Mother wrote that he had been born in 1868 (the same year she was born) and had died May 26, 1926, when he was fifty-eight years old and I was twenty-three.

My father didn't look after himself when he was working, any more than I did. He wouldn't have dreamt of coming in to get a coat if it turned cold or rainy, but—when I think about it—he wasn't much different than me, or all the rest of the Scottish farmers I'd grown up with. I was saddened, but knew I had to get on with my life, just as he would have done if it'd been me that died. Maybe it was because I was so far away, but later I discovered he spoke to me in my thoughts, just as he always had.

About this time, I became interested in Helen Ramsey, the daughter of my cockney friend. I remember her as the first girlfriend I had in Canada, though our relationship never became serious. She was still a schoolgirl, training in a business college. I just thought she was attractive and I enjoyed her company and the fun of her cockney family, that's all. Although both Helen's mother and little sister were born with a "harelip," their afflictions didn't impair their sense of humour. One day I laughed with surprise when I saw Helen's little sister amuse the younger children by putting a pea up her nose and bringing it out through her mouth.

I liked the Ramseys and—after I moved on—I wrote to them for quite a while. I was happy to learn Helen's sister later had a series of operations in Montreal, closing her harelip. Helen herself was happy and had a job by the time I left.

There was also an Irish girl who gave me a rush. She was curly-haired and pretty, but she came on too strong. She was wanting a man, wanting to be married. Christ, I had no notion of getting married, neither then nor earlier in Dundrennan.

My sister Mary Twiname's three boarders—Taylor (an Irishman), me in the middle, and McLusky (another Scot).

It's true I'd had two or three girlfriends in Scotland, but nothing serious. Maybe I'd take a girl home from a dance, but then the next dance I'd take a different one home. I took what I could get, like most fellows, but I was anything but romantic. I liked

Scottish step-dancing, not this "pushin around stuff" I saw other fellows doing. As far as I was concerned, the Irish girl in Quebec was romantically way ahead of my thinking.

I knew by then, that women have the toughest time of all in life. They work longer, just as hard as the men, then they have the business of having babies as well. I didn't want to make anyone pregnant and, though condoms were available, birth control was not freely known about. Around this time I made a conscious decision. For finding my physical pleasure, it was safer to pay as you go for prostitutes, than it was to risk pregnancy and marriage. Marriage was definitely not part of my planning.

At that time, it seemed to me many men made the same choice, going to whores instead of forcing themselves on their girlfriends. Most men I knew went into Montreal, where there was a whole street of colourful whorehouses, licensed and government-controlled, with the girls regularly inspected and looked after by doctors.

Though I didn't know it at first, there was a "house" in Ste-Anne-de-Bellevue. This house made me wiser about the politics of the town, even though I never used it to satisfy my own needs. I was introduced to it through a motorcycle policeman, another friend of my sister's husband. One day, the policeman pulled up to me as I was walking down the main street.

He said, "Dave, are you busy?"

I said, "No. Why?"

He said, "I have a chore to do. The priest has had complaints from his women parishioners because their husbands spend too much time at the whorehouse. Now he's after me to raid the place, but I need help. You could hold the gate open to cross the railroad track, so I can get out quickly, if necessary."

Curious, and thinking this could lead to some fun, I willingly hopped on board his motorcycle, and off we went. To my surprise, I discovered the whorehouse was just over the fence and across the railroad tracks from one of the pastures of Macdonald farm.

The policeman parked his bike down a ways from the house, so those inside would have no warning he was coming. He walked quietly up the path and I saw him enter the front door. I turned to walk back to open the railroad gates, and Jesus! That policeman was back beside me in no time, running hard and out of breath!

"Christ!" he said. I can't do a thing!" Not knowing why, I joined him in his mad rush, both of us hopping on his motorcycle, speeding out of there as fast as we could go. Red-faced and gasping, he said, "My boss was in there!"

Now, his boss was the mayor of the town, so he was right. There was no raiding he could do that day, without him losing his job. However, somehow the will of the town wives and the church must have prevailed.

A few days later, three women came out of the whorehouse hauling their luggage along with them. I saw them as I was ploughing with Helen's father and a couple of other fellows, in the adjoining field. The women were young and well-dressed, with skirts below their knees. They had bobbed hair, considered daring at the time. They also were staggering, very drunk, but giggling and full of fun. My friend hailed them, saying, "What's happened?"

They came over to the fence saying, "The buggers kicked us out!"

We started kidding back and forth with them, all of us enjoying the banter. Suddenly, to my surprise and with the laughter of all, one of them hitched her skirts, pulled down her pants and peed in the grass, right in front of us. As they began to walk away from us, old man Ramsey said almost sadly, "So they've thrown you out?"

One of them turned back to us, flounced her skirt and defiantly replied, "Oh, yes. But we've been thrown out of better places!"

In 1928, four years after my arrival at Macdonald College, I decided to leave Ste-Anne-de-Bellevue. I had successfully worked and enjoyed every job available to me on the farm. Of course, since I've been free to choose, if I don't enjoy a job I won't stick with it. I've never understood why anyone would continue for years at a job he didn't enjoy.

I left Ste-Anne-de-Bellevue because my health began to falter. My six-foot body, raised in the cool Scottish climate, just wouldn't let me get a good night's sleep during the heat and humidity of the Quebec summers. I got a case of bronchitis that wouldn't go away. My weight dropped from 180 pounds at my strongest, to 155 pounds. I looked like hell. Though I still have fond memories of my time in Ste-Anne-de-Bellevue, I knew if I stayed, I'd become seriously ill.

It was about this time I became friendly with an aging janitor at the farm, a man named Brown. He was a strange old man who set my life on a different direction, for a time. Brown appeared about eighty years old—though he never told me his age—yet he had the energy of a young man. He had a very young wife, in her twenties, and a young son six years old. Another of his sons, about my age, was from a previous marriage. The older son tried working with his father, but he couldn't maintain his father's pace. The son left Macdonald College, finding another job in a smelting plant sixty miles further north, in Deloro, Ontario. When the old man saw how my health was failing, he suggested I might be better off joining his son at Deloro.

"You'll be paid the same wages for working seven hours a day at the smelter, as you are for working nine or ten hours here," he said.

This sounded pretty good to me, so I wrote his son, applied to the Deloro Smelting and Refining Company, gave my required month's notice to Macdonald College, and left without further ado.

Though I had feelings of regret about leaving Ste-Anne-de-Bellevue, they were mingled with excitement about the adventure ahead. I had the optimism of youth and few responsibilities except for my own interests. It didn't take long for me to discover, my leaving was a serious mistake.

CHAPTER FOUR

A Detour through Hell

We cam na here to view your works
In hopes to be mair wise,
But only, lest we gang to hell,
It may be nae surprise.

—Robert Burns,
"Written on the Window of the Inn at Carron"

In the spring of 1927, I travelled north by passenger train through Ontario forests still piled with snow between the evergreens, yet on the sides of the tracks I could see green shoots poking through here and there. I was so optimistic, I discounted the fact my new direction was taking me off the path leading to my goal of becoming a farm manager. I had been thinking only of the promised increased money on payday.

I stepped from the train in Marmora, the closest railway station to my new job. There I was picked up, along with others, and transported to Deloro, a village owned and managed by the Deloro Smelting and Refining Company. This experience haunted me for years to come.

I'd been told the name "Deloro" meant "Valley of Gold,"[1] a name given during the time when twenty-five mine shafts in the area produced $300,000 worth of this gold plus valuable arsenic.[2] Before 1900, a mill was built there, the first in Canada to extract gold from ores by leaching them with cyanide.[3] After the gold was removed, the remaining material was roasted to remove arsenic. By 1907, because this place was already equipped to handle arsenic, the Deloro Mining and Reduction Company was contracted to process silver ores mined in Cobalt, Ontario. In 1914, a year before its name was changed to the Deloro Smelting and Refining Company, the plant produced the first commercially

Notes to chapter 4 are on pages 239–40.

viable cobalt metal in the world, a metal worth even more than gold at that time.[4] But on the day I stepped down in Deloro, thoughts of gold became quickly tarnished.

As we neared the place, the green of the countryside ended. Everything appeared in shades of grey and the country quiet was filled with the ear-damaging noise of the machinery. *Jesus,* I thought, *this place is ruined. How can such ugliness be?*

Later, I learned the complete reason for the destruction, though at first, the tall, smoke-belching chimneys dominating the sky providing a clue. (The combined assault of years of cyanide leaching and the burying of crude arsenic waste in the early years, then more recently the continued depositing of tons of white residue—composed of 60 percent lime and 40 percent arsenic—ruined both the land and the Moira river in Deloro.)[5]

As we entered the town, I passed rundown buildings, paint smeared grey by emissions from the plant. A double row of about forty houses sagged together on either side of a road, bordered by dirty, melting snow and muddy ground. I recognized bunkhouses, a big warehouse, a company store, and a school. Close to the Moira River, several huge plant buildings squatted. In contrast with the beauty of Macdonald College farm, this place looked to me like a garbage dump.

In the plant buildings, ores were refined into pure substances like silver, or combinations like stellite (a very hard metal made of cobalt mixed with tungsten and chromium), as well as other metallic combinations. I'd been told that this place also produced arsenate of lead, used to spray weeds along railway tracks, and arsenate of lime, used to make "Paris Green," a pesticide used on potatoes. To me, pesticides were interesting because they were relatively new, rarely used in Canada until just four years before I went to Deloro.[6] This awful place was actually considered progressive, in its time.

At the smelting plant, I saw huge metal pots containing a repulsive-smelling liquid, bubbling and steaming. As condensation from the pots hit the ground, it melted into the late snow. The runoff looked slimy, leaching down into the mine area where I could see miners getting their drinking water. The more I saw of this town, the more certain I knew my move here was a mistake.

Company policy prevented unmarried persons from living in company houses on their own, so I boarded with the Browns, a family made up of the father—the son of the old janitor of Macdonald College—the son's wife, and his two sons. Most of the five hundred or so labourers lived in shared bunkhouse rooms provided by the smelter, but I had the best deal. Mrs. Brown came from Northumberland, in England, and Jesus! she made the best Yorkshire pudding. Served it with roast pork, every week. For me as well as all the single men in town, privacy was hard to come by, but we were used to that.

Of course, the smelter manager's home was the biggest and fanciest in town, befitting the fact that, as manager, he was also automatically the village reeve. His home was fronted with a lawn-bowling area, which everyone who belonged to the bowling club could use for enjoyment. Even I bowled there once or twice by invitation. There were also tennis courts and a tennis club. The manager lived in style, having a chauffeur to drive him through town, and several automobiles.

One of his company cars was an expensive McLaughlin Buick, a touring car about eight years old, with a top that folded down. He allowed the company baseball team to use it. This may seem a strange thing for the manager to do, but members of the Deloro Smelters were given first-class treatment because they were the village's chief source of entertainment and even fame. They played in the Trent Valley League, against the small towns of Madoc, Marmora, Tweed, and Havelock. The Deloro Mining and Smelting Company hired very talented amateur players, often university students, from Ontario and Quebec.[7] Though officially on the company payroll, baseball players were required to do very little work at the plant. The company's investment in the team paid off, because eventually, the Deloro baseball team won the Ontario Intermediate A Championship, then went on to the Senior Central Ontario League, challenging teams from Oshawa, Kingston, Peterborough, and Belleville.

However just before I arrived, it seems the baseball players were making a habit of getting drunk, damaging the boss's car when they drove it. In conversation, a garage man, who worked next to where I kept the horses, tipped me that the McLaughlin Buick was up for sale. He said, "You should buy it, because the boss is selling it for much less than it's worth. He just wants to end the worry of it."

I made an offer and bought the car for three hundred dollars, a very good price. I proudly polished and drove around in it for about a week, my very first automobile. But a week was all it took for me to realize I wouldn't need it for long. Other than the car and my pay, I knew there was little to hold me in Deloro, so I returned it and got my money back.

The only women in town were the wives of the office workers, foremen, police, and other company men. Though the married women organized bridge and musical evenings, Deloro had little to attract single women; not even prostitutes lived here. The closest whorehouse was in Marmora, about two and a half miles away. The three women in it were visited mostly by single men fortunate enough to arrange transportation. This wasn't difficult, because it seemed every married person in the town had a car, needed if they wanted to get around on holidays.

Deloro had little indoor plumbing, so I bathed in a tin tub, like most everyone else. With no sewers, many toilets had buckets attached. Saturday night was the time most people chose to dump the damned things into the surrounding fields, digging the stinking contents into the ground. The overwhelming smell of the Saturday night bucket brigade in Deloro is a powerful memory, one of many I wish I could forget.

As I got to know the people of the village, they seemed congenial enough. I knew the long-term residents must have been tough, struggling in years past through plant closures and hard economic times without any government assistance. But it seemed to me they had a strange way of expressing themselves, sort of "out of it" or "bushed." Sometimes they didn't even recognize how funny they were. The local newspaper, The Deloro

Once-A-Week provided examples of their humour, but it was the people themselves who entertained me the most.[8]

Deloro Smelting & Refining Co. Ltd., circa 1912

One day I hitched a ride on a cart driven by one of the local farmers who delivered wood to the houses. We commented about spring being a poor season for transport on village roads. Our horse-drawn wagons were fitted with sleigh runners on snow or ice, then refitted with wheels on hard-packed ground. During this particular northern Ontario spring, road surfaces were changing from mud to ice, then back again in no time at all. Neither runners nor wheels work in mud, and soft mud was our problem this day.

As we rode slowly behind the struggling team, we passed a farm woman, a customer who called out to my local acquaintance, "John, when will you be coming?"

Dead serious, he called back, "Don't worry. If it stiffens up in the night, I'll slip it in in the morning." He didn't qualify what "it" was, but my randy mind quickly provided its own interpretation. I started laughing so hard tears came into my eyes. The driver and his customer just stared at me as if I was crazy, neither one realizing the joke they had created.

Another time it was the policeman who made me laugh, but not quite so loud and a little more carefully, to be sure. He was a big fellow named O'Neill, who revelled in his tough-guy reputation, yet he could be tricked with only a few words said at the right time.

It all happened because a fellow named Johnson and I started tussling in fun, as we waited in the gathering at the carpenter's shop, where we regularly received our morning work orders. I gave him a nudge and Johnson slipped on the ice, falling to the ground and spinning around. Everybody laughed, but one of the other men decided to use the incident to bait O'Neill.

He went to the policeman and said, "You know that Caldow? He's one tough bugger. Strong, too. I just saw him pick big Johnson up, toss him, and spin him around." This was a bloody lie of course, but O'Neill took the bait and came looking for me later.

On my way home he sidled up and said, "Caldow! I hear you're a bit of a wrestler. What about if you and I have a go some time?"

I said, "If you like. Where?"

"How about if we use your horse stall?"

I nodded in response.

"But you'd better bed it down with some straw, first," O'Neill said.

Up close and quiet I said to him,"If you want it bedded down with straw, you'd best find some. I have no need of straw because I don't intend to be hitting the ground. If you need padding, get it yourself."

"I'll get back to you," he said, walking away.

But I never heard any more from him. I guess he found out I was a bigger bullshitter than he was.

For men, year-round entertainment consisted of billiards, in the Single Men's Recreation Club, also called "The Hub." It had eight billiard tables, plus poker. On pay night, poker was played for big money. When the plant manager bet, the kitty was hundreds of dollars, a fortune to me. I watched, but playing poker for myself was of no interest.

I'd been turned off it since I was fourteen, when I was challenged to play poker by the hired men back in Scotland. I'd gambled and lost the few shillings I'd earned snaring a hare and selling it. This loss killed any joy in the game for me. Besides, if the loss hadn't killed it, my father would have killed me, if he'd found out. I made a vow then, "No more bloody poker for me!" So in Deloro, I'd watch the betting until I got bored.

It was boredom helped lengthen my already long work day. To fill time, I worked an extra job from six to eleven in the evening, dumping ore into the ball mill, a machine that crushed the ore to powder before it began the smelting process. For hours I dumped ore sacks, about two feet by ten or twelve inches round, weighing about a hundred pounds each. Very heavy work as the hours wore on.

My regular job was driving and caring for a team of big Clydesdales, hauling heavy loads with a large work wagon. Sometimes we transported gravel, or limestone required for the blast furnace. Often in two-in-the-morning darkness, accompanied by a security guard for protection against theft, I hauled up to three loads of numbered, insured silver bars down to the Marmora train station for shipment to London, England. Every day, I'd to feed, brush, and otherwise care for the Clydesdales and their equipment. The harsh demands of my new job in Deloro quickly eliminated my dream of a healthier existence, of having the same pay for shorter hours.

It surprised me that the people who lived here didn't mind having the smelter for a neighbour. It proved profitable for most of them. Nobody anywhere worried much about pollution in those days. Some farmers in the surrounding area had gravel pits, making money selling gravel to the smelter. I was beginning to learn, neither the people nor the smelting company were all bad, like most things in life.

The company, in its own way, took care of us to some extent. Our pay was pretty good and, because the office organized the twenty-five cents a month deductions, anyone who needed it had access to a sizable fund for needy cases, with money dispensed by a five-person relief committee (three from hourly rated employees and two from staff). There was a company doctor available to residents of both Marmora and Deloro, paid for by a dollar-a-month deduction from single men and a dollar twenty-five each month from married. British subjects, like me and the local farm labourers, all could find extra work at so-called easy jobs assigned to us. In comparison to the work given "Europeans," our jobs were damned easy, believe me.

The hundreds of isolated, homesick, and probably poisoned Europeans in Deloro were legally contracted in Montreal, then made to work in the most dangerous jobs in the whole polluted village. The company claimed they always had three crews at the ready—one crew to work, one on the way up, one crew recruited and waiting for their turn.

These were the men who retrieved the arsenic. Members of some work crews spread the thirty tons per day of crude arsenic over the hearths of the reverberatory furnaces.[8] Other crews cleaned the hearths, removing the residue after much of the arsenic was heated into gases and smoke. (This poison-laden smoke—initially with temperatures at or above 295°C—was then forced through three massive chambers, each divided into thirty-nine condensing "kitchens,"[9] where heavier dust particles settled into different grades of arsenic. As the smoke left the last "kitchen"—at the rate of 18,000 cubic feet per minute—it carried with it extra fine arsenic dust that would not settle out. To capture the last of the arsenic dust, the smoke—with a temperature of 90-100°C—was fan-forced into the "bag house.") This was the worst place of all to work, but it wasn't just the heat that made it bad.

The bag house held about two hundred circular woollen bags, each thirty feet long and eighteen inches in diameter. The fibres in the woollen bags collected the extra fine arsenic dust from the smoke as it was forced through them. These bags were mechanically shaken into hoppers once per shift, then workers removed the arsenic from the hoppers.

In spite of the heat, the Europeans working in the bag house covered themselves from head to foot with clothing, tying collars tightly around their necks, rubbing a brown, protective paste—provided by the company—on their exposed skin. They also wore breathing apparatus.

These crews were told, "Work slowly, be careful not to work up a sweat."

In spite of the clothing precautions, the protective paste, and slow pace, the extra fine dust filtered through, collecting on the men anywhere there was moisture—their eyes, the sides of their mouths, their genitals.

Furthermore, the dust sometimes contained more than arsenic, as if arsenic wasn't enough. (Arsenic can be a volatile gas that condenses into a fume that can coat everything.) When pesticides—such as arsenate of lead—were being manufactured, other components probably were added to the mix.[10] Caustic soda, lead, and nitric acid could have been included in the process, producing a dust not only poisonous, but also caustic.

No wonder the Europeans, as they came off shift, were made to shower carefully and told, "Put on more brown paste to burn the arsenic out of the damp places."

In the showers one day, I watched the bag house men jump and hold on with pain as the brown cream touched their skin. I could see they were faced with a terrible choice between the immediate pain of the cream or the even greater pain of the deep burn that would surely follow from their exposure to untreated smoke residues. I couldn't believe what my eyes were seeing. Naked, except for the brown paste, they looked like pictures I'd seen of Indians in war paint, but in Deloro the war paint was painful. The sight of the men in the showers made my skin crawl with horror.

From then on, I couldn't forget them working in that foul smoke and dust. I wondered why the Europeans didn't all leave. Then someone explained that even if they saved enough, the company wouldn't allow them to pay for their transportation in a lump sum. Only small payments were accepted, keeping the men working longer. If they ran, O'Neill and the RCMP brought them back because they had broken their contracts. I heard of only one man who made it clear away. I was told he ran far enough to reach a tall cornfield, where he hid until it got dark. Try as they would, they couldn't find him.

Even though there was a "safety committee," it didn't take me long to figure out that company attempts at "protection" weren't enough for these men.[11] Many workers looked weak, pale, with swollen bellies too big for their sickly bodies. It was an eerie feeling when I realized these people reminded me of the Scottish horses I'd seen treated with arsenic when they had worms. At that time, arsenic was considered a conditioner for horses, but big Clydesdales were given only as much of it as could sit on a ten-cent piece, that's all. I remembered my father telling me that the relatively tiny dose was the reason the horses appeared fat, with swollen bellies, just like many of the poor workers in Deloro.

As I became fully aware of the potential for horror there in that smelter village, I was convinced arsenic was seeping through every crook and cranny in the place. Proof came when I cut myself shaving and got the damned arsenic dust in it, sending me to the makeshift company hospital for treatment. There I saw a man, one of the Europeans, who'd had his genitals almost burned away. The nurse explained that poisoned dust had caused the burn because this poor fellow didn't know enough English to learn how to properly apply the protective paste.

Seeing this man in his agony, knowing the draught horses I'd given the best of care were sick with something similar to bronchitis in humans (called "heaves" in horses), I made up my mind. No matter that mining paid more—give me a farm any day. I'd paid for my own transportation, so I had no contract to break. I got myself out of Deloro and on a train for Montreal. Mining and smelting were not for me.

As I rode back through those Northern Ontario forests, further and further from Deloro, my spirits lifted. I planned to go West . . . see the rest of Canada . . . work the farms along the way.

Maybe I'll just keep on going, all the way to Australia, I thought.

But I was changed by this town, made older—maybe stronger—maybe more capable of accepting the challenges to come, on the Canadian prairies. The images of Deloro and its workers are etched in my memory. In spite of its golden history, to me the name "Deloro" means the saddest place I've ever been. A true hell on earth.

CHAPTER FIVE

Claresholm

Ae night the storm the steeples rocked
Poor labour sweet in sleep was locked . . .
—Robert Burns, "A Winter Night"

Within four days of my return to Ste-Anne-de-Bellevue, I found myself paying the fifteen-dollar fare to board a train called the Harvest Excursion, travelling west to Winnipeg, Manitoba, looking for work. Along with me came Helen Ramsey's brother Leonard, a fellow named Bob Hogg, and an Irishman whose name I don't remember—all of us heading into an unknown future in the vastness of the Canadian prairies.

Even the train ride held surprises. Because I liked to mind my own business, I would never have predicted I'd soon be commandeered to help the RCMP. I became their aide, all the way from Fort William, Ontario, to Claresholm, Alberta. I let the Mounties involve me because I couldn't stand to see Helen's brother punished for something he didn't do.

The boredom and confinement of the long train trip across the monotonously flat prairies caused some of the many Frenchmen on the train to get too rough and careless at the station stops. With just two Mounties on board, the Frenchmen couldn't be controlled. At Fort William, Ramsey and I heard a loud crash followed by tinkling glass. Then we saw a Frenchman throw an empty beer bottle, aiming at a post but hitting a person, bruising and cutting him.

We both yelled at the fellow who threw the bottle, but Ramsey told me, "I'm reporting this to the Mounties. Why aren't the police here where the trouble is? Why aren't they doing their job?"

Notes to chapter 5 are on page 240.

I said, "Leonard, keep your mouth shut. It's not our business." But he wouldn't listen.

He went straight to the Mounties. Christ, they took Ramsey to the back of the train for questioning. By the time the Mounties let me talk to them, they were convinced poor Ramsey was as guilty as the guy who threw the bottle! In spite of my desire to keep to myself, before we arrived in Winnipeg (as far as our fifteen-dollar fare would take us) I convinced the police to release Ramsey into my custody. In return, they made me promise to report his whereabouts until we reached the end of the trip.

On our arrival at Winnipeg station, there were no people looking to hire us. We milled around for a bit, uncertain, with me finally deciding what to do. "I'm going as far as the Excursion will take me," I said. "I can pay the cent a mile they charge to continue to Calgary. I want to see as much of Canada as I can." The others took a vote and decided to go where I was going.

Dutifully, I let the Mounties know our decision, but I really resented having to report to them. They were so unfair to Ramsey.

Compared to Winnipeg, Calgary station was a noisy circus. Recruiters of all shapes and sizes called to us, surrounding us, pointing to their hats sporting brightly coloured bands naming towns needing men for the harvest. Each enticed us to come to their town.

I saw the name "Claresholm" and liked the sound of it. I said, "Claresholm's where I'm going!" My companions followed suit because they'd heard somewhere this town was in good wheat country. Of course before leaving the train station, I again had to report Ramsey's destination to the Mounties, all because I helped a friend out of a jam.

We caught the train to Claresholm in a heavy downpour that continued for a week after our arrival. Yet, in sharp contrast to Deloro, I felt comfortable and liked Clareshom even before the clouds lifted. When the sun shone, I was impressed with this town of about fifteen hundred people. Its grain elevators and painted buildings stood tall and proud above the surrounding green, golden, and flat farm fields. It was all the more beautiful because the rainbow-decorated foothills of the Rockies were its backdrop. We learned the town's tall trees were each planted by human hand, with only prairie grasses naturally nurtured there. Nature also provided a slough, making this spot a rare wet place on the dry prairie. The Canadian Pacific Railroad needed a watering hole for trains on their way to Calgary and Fort McLeod. Thus, in 1891, the gutsy little whistle-stop town of Claresholm was born and endured.[1] During that first week of rain, I learned the town history, but I didn't find a job.

No farmers were hiring. Why would they pay wages for hired hands huddling indoors? But after the rain stopped the action began, and Ramsey was the first hired, to drive a team of horses pulling a road grader. Then Hogg got a job in a creamery and the Irishman went "stooking." Though my friends found work before me, I wasn't worried. I still had four hundred dollars in my moneybelt—saved from my jobs back east—even though I'd bankrolled forty dollars each to the others for food and train fare from Winnipeg.

Though my friends knew I wasn't destitute, both Hogg and Ramsey must have felt sorry for me when they left me behind in town. They quickly repaid the money they owed me, without me having to ask. But the Irishman had no such concerns or intentions.

I never saw that Irishman again, even though eventually I went looking for him to get the forty dollars he owed me. He hid from me. Eventually he hopped a train to avoid paying me. I felt angry and hateful, feeling betrayed by someone who had been a friend. No wonder I've forgotten his name. It wasn't the loss of the money that bothered me, it was his dishonesty. While I was dealing with this Irishman I thought often of my father, realizing honesty was now as important to me as it was to him.

While waiting for a paying job, I volunteered to help Claresholm's blacksmith, a man named Alex Hutchison.[2] I removed the worn horseshoes from the horses he reshod. As I watched him work, I thought Alex was the biggest man I'd ever seen, and very powerful. Yet he was kindly, known for always looking after people who needed help.

Though he was older, Alex and I had much in common. Alex had come from Scotland to Canada in 1913, when he was twenty-one, like me. Like me, he'd learned his trade from his father. As farming was for me, blacksmithing had been for his family, with several generations following the trade. He was over six feet tall because size ran in his family. He had a twin sister, taller than my six-foot-tall Aunt Mary. I found him to be a good companion, able to talk about the community affairs, sports of all kinds, and of course his interest in horses, especially racehorses.

Alex must have felt an easy kinship to me also, because as he got to know me, he offered me a full-time job at his forge. I told him honestly, I appreciated the offer, but my heart was in farming. I wanted to work on the harvest. To help me, Alex recommended me for a job on the Walker brothers' farm, paying $10 a day for general farm labour; $12 a day when driving a four-horse team . . . good money in those days. Alex remained a valued friend, even though one day Teddy Mack, the CPR station agent and talented town joker, used me to trick him.

Teddy told me, "Dave, find Alex and tell him to get over to the station and look after his friends who've just arrived, asking for him." I did as Teddy asked, then followed Alex as he ran excitedly to the station, all the way wondering who it could be that had come into town. With a grin, Teddy introduced Alex to his "friends," two monkeys being shipped as freight in a cage. Alex didn't mind the guffaws that greeted his surprise, because he was a good sport who laughed at the joke just as hard as the rest of us spectators.

Teddy Mack was born in Ontario, a true Canadian, talented with the ability to mimic any accent. When I met him, he had been in Claresholm for twelve years, using spare moments to practise his mimicry of the world-wide accents of coal miners who came off the trains. He entertained in local operettas and plays. I had many laughs observing him, especially one time when his wife was away on a trip. Teddy was on a tear with some Scotsmen, new in town. They were buying him drinks because they really believed they'd found a kinsman. Teddy's broad Scottish accent was completely convincing, though his family had been Canadian for several generations. Yet, Teddy wasn't just a funny man. He genuinely enjoyed people and was concerned about the community of Claresholm. He served as a town councillor and later, I learned, he served ten years as mayor of the town. He would have been an entertaining mayor, that's for certain.

Scottish dancing was another form of entertainment in Claresholm, sponsored by the St. Andrews Society. It was fun, but it got me into a bit of trouble with an English girl who had also come out West to help with the harvest. Her name was Clara Chambers, from Bolton. She was a nice enough girl, fun as a dance partner, but too damned serious about romance for my liking. One night, after we'd had an especially good time enjoying a visiting Montreal group playing Scottish tunes and demonstrating Irish step-dancing, she started telling me how much she cared about me.

Never one to beat around the bush, I told Clara damn straight, "I'm not one to be getting married now, girl. There's plenty of interesting places in this world I want to see before I settle down. Marriage is the last thing on my mind right now, and you can be damn sure I won't change my mind, either." But she wouldn't be put off easily and it made for some discomfort, later on. I did enjoy dancing with her, though.

The Claresholm beer parlour turkey shoot was almost as much fun as the dancing, with the added advantage of not having romantic complications to spoil it. Most of my teetotalling religious friends said they didn't want to enter this contest because of a rule stating a drink was mandatory before entering. Just for the hell of it, I challenged and embarrassed them into giving it a shot. With them giddy from the drinks and me teasing them, I won more turkeys than I'd ever expected. We all ate like kings after the contest, sharing my winnings. It was a good thing there was plenty of entertainment in town, because life on the farm where I worked was often trying.

I remember my employers—the Walker brothers—as a strange pair, so different you'd have never known they were brothers. Gordon, at that time unmarried, had been a member of Parliament for almost four years. He had his BA degree from the Manitoba Agricultural College, and he had been an instructor of animal husbandry and farm management at the Claresholm School of Agriculture, until his election. When Gordon was at the farm, he was usually happy and optimistic around me. Gordon didn't own the farm property, he had rented it, then convinced his married brother Bert to come from his own farm in Antler, Saskatchewan, in order to make the Claresholm spread a success.

When I was there, brother Bert didn't seem very happy, often moaning and complaining. As a bachelor myself, I sometimes wondered if his marriage was the reason for Bert's miserable nature. Later I decided he was probably jealous of Gordon's many successes, the usual rivalry of brothers. My father's training helped me to impress Gordon with the quality of my work, but Bert was never satisfied.

My job included watering eighty head of horses, driving a six-horse team for planting and other heavy work, milking cows, feeding calves, caring for Gordon's prized Bark Rock chickens, and fattening pigs for market.

The speed I fattened the pigs impressed Gordon the most. Normally, they had taken a year to get their pigs ready, letting them run free on five acres seeded with grain. Those pigs might as well have been living on a race track, the way they ran around. As my father would have done, I penned the pigs with clean straw, feeding them grain and milk from the cows, and they were fattened before November. Gordon was happy because they got their pig money earlier.

In the winter, before the snow arrived, I had to bring the horses to water, all alone, except for my well-trained cow pony. Without the field work, there was time for learning new skills. I'd always wanted to learn lasso, so I began lassoing everything as I trailed the horses—fence posts, even the dog. One poor old mare was always last in line because the side bones on her front feet were sore. The devil in me got to thinking, maybe I could lasso this mare, thinking she'd be easy maybe, because she was so docile.

I tied the end of my rope to the saddle horn, whirled my arm around in lazy circles, then let go at just the right moment to loop the lasso over her head. As the rope settled over her neck and the lasso tightened, I got the surprise of my life!

Jesus! That mare took off. She bolted so hard, she broke the work-worn cinch on my saddle. When the cinch separated, I was pulled off my horse and dragged behind the mare for a couple hundred yards—still on top of the saddle, holding onto the saddle horn! Eventually, that mare came to a stop, leaving me shaken and feeling foolish, but unhurt.

My old cow pony didn't move from where we'd parted company. He just stood there looking at me, as if he was thinking, "You bloody greenhorn!" If he was thinking that, he was right! He was a smart old horse.

Though that mare defied me, I would never have thought of beating her into submission. I'd learned from my father in Scotland, gentleness is needed to tame a horse. But as in Scotland, I learned that here in Canada, there was some who didn't agree with that belief.

One day in town, I happened across one of the local cowboys, kicking a horse in the belly with his big cowboy boots. Though the fella was bigger than me, I couldn't stand to watch. The horse had terror in its eyes.

I yelled, "You kick that horse again and I'll kick you right back, in your belly!"

Luckily, he only glared at me in response, but I knew I'd made an enemy, right there. Even though the town was small and I couldn't avoid seeing him, I didn't care. He never spoke to me again, but that never bothered me. I didn't need to speak to the likes of him.

When the deep snows of winter came to Claresholm, life on the farm became relatively easy. The fella who ran the separator even had time to play dentist, as a favour for me. I knew he owned a pair of dental forceps, used for putting screws in difficult places

on the separator. I decided to ask him to help me end a toothache
in my lower jaw that was driving me crazy. He agreed to try and
he did his best. I could feel him pulling and it hurt some, but not
enough to stop me giving him directions. Eventually, he couldn't
take any more of it, so he quit, leaving me with a bloody mouth
and an even more painful tooth.

It was then I went to Arch McGregor, the local dentist. I knew
him because he was president of the St. Andrews Society. He finished the pulling of my tooth and gained a funny story, to boot.
Ye see, the chair he sat me in was next to a leaking gas heater that
smelled like a rotten drain. Just the smell was enough to make me
woozy—putrid. I can still smell it now, as I think of it.

As soon as McGregor got my tooth out, he asked me for pay-
ment. I handed him a twenty-dollar bill, but he had no change. He
left me sitting in those gas fumes as he rummaged in his back
room, looking for smaller bills. By the time he came back, I'd
fainted, overcome from breathing the leaking gas. But that was-
n't the way he told it. From then on, McGregor told his joke
about this Scotch man (me) who fainted because he thought his
dentist had run away with his twenty-dollar bill.

In the deep snows of winter, animals not slaughtered and sold still
had to be tended, though most outside jobs became impossible.
Almost all prairie farmers I knew liked it that way. They'd read,
curl, go target shooting, or travel to warmer climates.

Not me! I still had to stoke the fire in the stove that melted the
water in the thirty-foot-long watering trough, then I'd feed and
water all the animals . . . but I didn't stop at just the jobs I had to
do. I enjoyed working in the farm's blacksmith shop, fixing and
cleaning all the equipment. I'd be sharpening the harrow or cultivator teeth, or grinding the harrow disks. The biggest harrow
disk was twenty-one feet round, huge for that time. (Recently, I
saw one a hundred feet round, made for a farm in Texas—where
else?) Hell, I couldn't just lay around doing nothing or I'd have
been an old man in no time!

Some of the other farmers told me they couldn't understand the way I liked to work, but they didn't know I'd been working this hard all my life. I think my energy and my drive to work hard was what saved me during the depression, when money was scarce.

The Walkers were hit by the depression, just like everyone else, though it took a while for the size of the financial disaster to be realized in such an isolated prairie town. The news came to us first by radio, then bits and pieces of news about the other parts of Canada came from people who were travelling through town, stopping off from the trains. As the year progressed, the Walkers had difficulty paying me with cash. To help me, Gordon kindly allowed me to sell extra cream and eggs and keep the money. I worked a deal with Bolivar hatchery that resulted in money for me and help for us all.

Because Bolivar's wanted my eggs for hatching chicks, they gave me permission to put six of their roosters in with Gordon's prize Bark Rock hens. Bolivar's provided not only chicken feed and cod liver oil, they also paid me eighty-six cents a dozen—instead of the usual thirty cents—because these eggs were considered very desirable.

I needed my cream and egg money to pay the Walkers for the room and board of a friend, Walter Craig, whom I allowed to stay with me throughout the winter because he was in a difficult situation. Walter had been recruited and contracted in Scotland to work for a year on a farm in Manitoba, thus earning his fare to Canada. For his efforts he got his clothes washed, plus his room and board, but very little money. When his contract ended, he came to Claresholm looking for work. He worked for a time on a farm near the Walkers', where he met Bessie Foggo, the niece of Alex Hutchinson. Bessie, who eventually married Walter in 1939, introduced me to him at one of the Scottish dances.

One day, just as the winter of 1929 was setting in, I saw Walter in town looking so forlorn.

"I'm broke," he told me. "I've no hope of a job, and I don't know how I'll survive until spring."

I knew him to be a good little fellow, not hard to get along with, so I said, "Walter, you come home with me and we'll see what we can find." In return, he offered to cook for me, but a few days later he told me he'd never last in the kitchen.

"I'm damn glad you made that decision," I said, relieved, "because you sure can't cook." We laughed and continued our original arrangement until the Spring, when the Walkers hired him on my recommendation to drive tractor.

People feel good about the good things they do, and I feel good about having helped Walter. He appreciated it, too. Thirty-five years later he phoned me, telling me he and Bessie had moved to British Columbia and were living on Scott Road in Surrey, not far from where I'd settled. We were lifelong friends.

Bert Walker, however, was not such a friend. When Gordon left to go electioneering or working in the legislature in Edmonton, Bert became boss. His high-handed holier-than-thou attitude and his moaning was beginning to drive me nuts! He often bungled simple jobs, so even his wife would sometimes say, "It's too bad Gordon isn't here," when she wanted something done.

That's how the pig incident started. We were sitting at the table when Bert's wife said, "It's too bad we didn't kill the pig before Gordon left." She didn't mean it as an insult—it was Gordon who customarily killed the pigs.

Her words set Bert off. Angrily he said, "I can kill the bloody pig! I used to kill our pigs back in Antler!" Then he ordered me to help him, so we both walked out to the sty.

He told me to feed the pig some oats to keep it still. I did, then Bert tried to shoot it, three times: once in the side of its head, once too high on the top of its head and the third time through its mouth aiming up through its nose. After each try, blood splattered all over my clothes, my hands, and my face. The poor bloody animal squealed, struggled free, and ran around terrified. Each time I'd have to catch it and wrestle it calm enough for Bert to aim again. All the while I was trying to make damn sure my own body was out of his line of fire.

Not one of Bert's shots either hit me or killed the pig. Eventually, the animal just stood up shakily and, rather calmly, walked away. I told Bert he'd better leave it be, because the meat would never cure with all that running around, so he let it go. Six weeks later I shipped the pig to the butcher, who couldn't even tell that pig had been shot! Pigs are tough, but Bert was very inept. I was disgusted.

My disgust grew stronger another time, when I watched Bert make himself feel big at the expense of a greenhorn Englishman who had come to Claresholm via Quebec. I was driving a six-horse team, seeding for the fall harvest. The Englishman was an extra hand they'd hired to help me. Instead of demonstrating the correct way to do things, Bert just handed the fellow a seed drill and told him to follow me. When he didn't put the drill in the ground correctly, Bert laughed and made fun of him. I stood it for a while, then I climbed down off the machine and gave Bert what for!

"What's the matter with you?" I said. "What are you there for? Can't you show him what to do?" Bert went red in the face and stomped away, leaving me to teach the new man. For sure, working with Bert was less and less enjoyable as time went on, so I had leaving on my mind even before my sister Mary and her husband Tom Twiname came for a visit.

Since I'd last seen them, they had moved to the Kelowna area from Quebec, and my brother Bob had joined them in Kelowna when he came to Canada. Tom had asthma, so Mary encouraged him to move to the Okanagan because she knew the dry climate would be good for him. Both Tom and Mary were worried about me because I again looked too thin.

I had been sick, with my stomach bothering me off and on ever since I left Deloro. I'd been to the doctor in Claresholm, but he couldn't find the problem. Eventually I became so rundown and nervous I was afraid to work the horses, almost afraid to walk up to them. My illness completely destroyed my confidence for a time, and Bert's carping didn't help me either.

In 1929, I knew that leaving Deloro for Claresholm had been one of my better decisions, but I also knew it was now time to move on. I decided to say goodbye to the Walkers' farm and to the people of the town, but not to Clara Chambers, the English girl. She never did believe I'd said no to marrying her, so when I left, I left without telling her.

Whenever I think of Claresholm, it's the goodness of most people that remains in my memory. They had courage to build and maintain that prairie town, standing tall on flat plains, close to the rolling foothills of the Rocky Mountains. It was with mixed feelings I left for British Columbia, though I was glad to accept the offer of help from my family. I was hoping to recover my health, find work, and survive the Great Depression.

CHAPTER SIX

The Okanagan Valley—Surviving Together

My mirth and gude humour are coin in my pouch,
And my Freedom's my lairdship nae monarch dare touch.
—Robert Burns, "Contented Wi' Little"

From Calgary I caught the train to the West to Kelowna, located in the Okanagan and Kalamalka Lake country. It was spring so Kelowna was all in bloom. Here, I lived with my sister Mary and Tom, during the considerable time I spent taking medical tests, searching for the cause of my severe bloating and pain. My brother Bob had recently married, and his wife was also called Mary. In between bouts of sickness, I helped Bob with his orchard job. Eventually, my doctor decided my ailments were caused by my appendix, so he removed it and I began to feel much better.

About 1932, in the heart of the depression, Bob lost his job when the market for apples collapsed. Though it was a disaster, we all survived because we weren't living in town. We could get food. If a deer run past us, we shot it and ate it. We had to. With my health improved, I accepted an offer that helped me as well as Bob and his family survive the hungry thirties, when there were no paying jobs available, when almost no one could buy farm products, pay rent, or purchase anything requiring money.

The offer came from a man named A.S. McArthur, who managed the "Save Store" in Salmon Arm, where we dealt when we lived just south of there.[1] Like Kelowna, Salmon Arm is located on a beautiful lake—the Shuswap—with a look to it that reminded me of lakes in Scotland. McArthur owned a three-quarter section of farmland with a house on it—in Tappen—located fifteen miles from Salmon Arm.

Note to chapter 6 is on page 240.

The land near Tappen had been prime beaver country in the early and mid-1800s. By the time we arrived, the small village was well-established, built close to the CPR, which connected it to the trade routes of the world. There was no trans-Canada Highway, so trains transported all freight and shipped the farmers' produce, as well as moving to market all the trees felled in the local logging industry. There were a few crank-handled, wall-mounted telephones in use and some electricity. The town had a post office in the Tappen Co-op.

The Co-op had been established by the Tappen Farmers' Exchange in 1915, in order to get a fair deal for the local farmers when marketing their produce or buying supplies and equipment for their farms. The Co-op enabled locals to buy their livestock feed in carload lots for reduced rates. Because the main source of power was still the horse, this Co-op even handled harness repairs. Eggs were graded here, then shipped to market in wooden crates. In the 1920s the store had been a success for the local farmers, up until the Depression slump hit here, as it did everywhere. Neither the Co-op nor McArthur's Salmon Arm stores were exempt from the devastating effects of the slump. This was the reason McArthur had made us his offer.

His rental place had some cleared land, but it was mostly bush, and he stood to lose it all to taxes if it couldn't be made profitable. McArthur offered us free rent for the clearing of it. Because it was now fall, we weren't very keen on accepting, at first. We knew there was no hope of seeing any profit or trade goods 'til at least the spring. We didn't know how we would buy food or fuel in the meantime. Yet every time we wrote to refuse McArthur's offers, by return mail he'd up the ante for our acceptance. I kept every one of his letters, which later proved to our advantage.

In his replies, McArthur offered $25 worth of groceries, at that time enough to carry us to spring, plus a supply of coal we could use when blacksmithing for the stock. (Blacksmith's coal is specialized to give very high heat, so you can't use it in a closed burner. It is so full of gas it will blow up. Bellows have to be used for blowing it to increase heat, instead of the draft opening you would use in a closed heater.) McArthur also wrote that he'd donate some wild horses he was holding on his other farm, on condition we break them. He also promised to supply twenty-five milk cows in spring, and allow us to keep the money for the cream from eleven of them. He even agreed to supply blasting powder to clear the stumps from the land, on which we hoped to earn money by growing and selling grain. The more land we could clear, the better it would be for us and for McArthur. Eventually, both Bob and I could see we had a bargain to our advantage, so we agreed to meet with McArthur to close the deal.

He was as nice as pie when we first met. My brother Bob liked him, but I rarely agreed with Bob's judgment of people. Bob was a good man and he thought everybody else was good, too—but me, I was more suspicious. In spite of McArthur's jovial manner, something about him made me uneasy.

When I asked McArthur about creating a written agreement, he said, "Everybody knows me around here and my word is good. We won't need any written agreement."

When we left him I said to Bob, "Mind, we'll have to watch that guy." I went to a lawyer to be certain we had all we needed to protect ourselves, and the lawyer told me McArthur's letters could serve as an agreement because all his promises were written in them. Knowing that, I felt it was okay to go ahead with the job.

It was the end of November when I went to the Tappen farm, alone. Bob couldn't come right away, because Mary needed to stay put for a bit. About a month before, she had given birth to their second son Andy, a brother for three-year-old Bobby. I began by plowing the cleared part of McArthur's property, getting it ready for spring planting. I tidied the house as best I could, painting the inside. I also cut myself some wood for heat and cooking. Two months later—about the end of January with snow on the ground—Bob, Mary, and their boys joined me.

Soon, Bob and I set out for our first job together since we'd worked the farms in Scotland. Tappen's cold, crisp, clean air felt good because we were working so hard. The noise of our axes broke the winter silence as we felled trees and cut limbs from their trunks. Steam rose into the air from the heat of our bodies, and our arms ached as we pushed the crosscut saw blade back and forth, cutting the logs into firewood lengths. It felt good to be working with Bob again, stockpiling the wood we needed to stoke three stoves for heat and cooking. Eventually, we were both tired, stumbling along behind the horses as we hauled the wood back to the house on our horse-drawn work sled. Then more chopping and more stacking would end our day.

Imagine our disappointment when Mary said, "There's something wrong with this wood. It's hard to get the fire started, and when it does burn it doesn't give enough heat."

We learned the hard way, standing dry wood was good, but green wood just wouldn't burn. Even with dry wood that old house wasn't very warm, especially in the middle of the night and early morning when the embers were almost out. We were heating almost as much of the air outside as we were inside, because there was no such thing as insulation in them days. It was no small job to keep ourselves warm and supplied with hot food, I'll tell you.

Using the last of our savings from our job on the apple orchard, Bob and I bought a cow to have milk for the children. We bought lumber from a nearby sawmill to build a chicken house. At the start of spring, we stocked the coop with day-old chicks—two hundred of them—bought in Milner. We chose Bark Rocks for eating and Leghorns for producing eggs. At first, we hung big electric lights down low to keep the chicks warm. Our money wasn't wasted. The chicks became healthy chickens and fed us well.

In Spring, after the snow melted and the ground softened, we started clearing stumps from the remaining eighty acres, using McArthur's blasting powder. We had a plan we thought was safe for both of us, and—for quite some time—it worked well.

I held the powder sticks while Bob drilled the holes, but as he drilled, the holes filled with water from the melting snow. Nevertheless, I put the powder sticks in, then packed clay around them, to force the water out. I took a connecting fuse, long enough to reach all ten drilled stumps, and nicked it about every two inches. Then to it I attached the ten fuses from the placed dynamite sticks. Using this connection fuse, we could ignite all ten sticks, one after the other. After I lit the connecting fuse, we'd run like hell to a safe spot, at least two hundred yards away. We knew we only had a few moments before the first stick would blow, throwing stumps, mud, bushes, and rocks at us.

Our plan worked fine as we cleared the first fourteen acres. We toiled long hours hauling the exploded stumps and burning them. As soon as we had enough land cleared, I started the ploughing and Bob carried on with the stump removal. At harvest time, we knew we had a good chance to reap not only grain, but also perhaps, some money. The more we cleared, the more money we hoped to gain, so we both felt it was a big advantage having two men to do the work. At this time, we never dreamt we'd lose that two-man advantage.

In a relatively short time together, we had accomplished a great deal. As well as clearing the land and establishing the chickens, we bought some cows and we broke some of the wild horses McArthur sent, training them for plowing.

Breaking these horses was very different than the horse-breaking I'd done in Scotland. Being wild, these horses had not been used to people at all. They'd been running the range all their lives, and most of them weren't young. Added to this, some of them had been abused after they'd been caught. One of them tried to kick us whenever we'd walk towards him, because he thought he was going to be hit.

One day, when we were about to take the horses to spring pasture, I climbed up on my saddle horse and led another mare over for Bob to use. The mare wasn't a heavy horse and I thought she'd be okay for riding. She'd been working for us for some time, pulling and ploughing.

Maybe Bob knew something I didn't, because he said, "You get up on her back first, if she'll let you. Then I'll ride her."

That was fine by me, so I tossed the saddle over her back. When I tightened the cinch, the mare arched her back, giving me a hint of what was to come. I mustn't have been too worried, because I didn't go in and change into a solid pair of boots. I was wearing low rubber boots because the ground was very wet.

As Bob gave me a leg up, I said, "You hang on to that halter!"

Just as I was putting my feet into the stirrups, her back came up again. She made one leap and Bob threw the halter over her head towards me, as he got out of her way. Up she went again and landed us facing in the opposite direction. The third time she carried me up so high, I could see right over the barn, I swear.

My mind screamed, "I'm getting out of here!" and the mare helped me out. My rubber boots were pulled off in the stirrups, my body arched and I flew off her back, landing on the ground. Luckily, nothing was broken, but it could have been a different story if I had landed inches aside, one way or the other. There was metal machinery, harrow points, and discs parked around us, and I could have landed on them. Knowing the riding of the mare would take time, I brushed the mud off myself, then rode over and borrowed a big saddle horse from a neighbour.

Breaking McArthur's wild horses took patience and kindness, but we eventually turned them into useful animals. They helped us greatly with the planting of our crops.

Next winter, we decided to get a bull to breed the cows. Our neighbour gave us an opportunity to get a Red Poll dairy bull for the effort of hauling him home. It sounded good to us, but that was before we'd made the bull's acquaintance. The beast—and beast he was—was an intruder on our neighbour's beef ranch about twelve miles away. His original owner, a dairy farmer, had built no fence to keep him in. This bull had wandered, searching for amorous cows. The bull's owner could have been sued or jailed because a farmer is supposed to fence his land. His bull could have impregnated any of the cows he encountered, ruining the integrity of the breeding records. The bull's owner was lucky our friend the rancher was neighbourly, unlike a Hereford rancher nearby. If that bull had wandered onto the Hereford spread, that dairy farmer would have been in court.

Our friend said, "The bull's healthy and without horns. He weighs about fourteen hundred pounds—not big, as bulls go. You can have him free for hauling him away, but—I warn the both of you—it won't be easy. He's a bad bull, this one is."

Bob and I knew our neighbour meant this bull would deliberately go after us—kill us, if we let him. His viciousness was probably the reason the bull was left running free in the first place, but we needed him for breeding.

We told the dairy farmer, "We'll be up tomorrow to get him."

The next day we took a team, sleigh, saddle horse, chains, ropes, and a "humbug" or heavy cast iron nose ring with a spring in it to keep it closed. This humbug was supposed to give us a secure hold on the animal, but we might as well have left it home.

We arrived at the dairy ranch about eleven o'clock in the morning. I eagerly climbed on the saddle horse, nudging him forward toward the bull. I was trying to rope the beast. He pawed the ground, snorted, then . . . Christ! He went right for the saddle horse! In no time at all my arse was up in the air. The bull was underneath my horse, lifting it with his head! My horse was neighing in fear and that damned bull was stomping and snorting in triumph. Somehow I managed to get myself and the horse out of there, slamming the gate as we went. All that went through my mind was, *Good thing that bugger has no horns!*

After a pause for a conference, my brother and I decided to get the bull running, then direct him into the barn. This part of the plan worked. We yelled and hooted, shaking and swinging the chains to get him going, forcing him towards the barn, then grinning in triumph as we slammed the door on him.

Then we paused to figure a way to get into the barn safely, ourselves. We knew, if we went in the same door as the bull, he'd kill us both. This barn was dug into the hillside, the ground at the back up pretty near level with the roof. We decided to take ropes and the humbug with us, climb up onto the roof, then into the hayloft . . . and the plan worked. So far, so good.

Next, we swung ropes down from above—lassoing him—each of us getting a rope over his head. The bull's eyes followed my rope as I swung it into position, but I hooked it quickly around his neck, then cinched it around one of the ceiling joists. Bob did the same with his. By this time the bull was really wild!

I'll never forget his furious eyes searching us out through the dust he was raising. He was bellowing, snorting, stomping and running at us, charging the supports trying to reach us. If he'd more space, it would have made him happy to bring the whole damn barn down on us.

We waited until he came towards us. Then we tightened the ropes, both of us pulling against the lever of the ceiling joist, using all the strength we had in us, until his front feet were off the ground and we were choking him. In a few moments, he stopped struggling and we had him—we thought.

I tied off my rope and yelled at Bob, "Hold him. I'll go down and put the humbug in his nose. Once that's in, he won't move. Then I'll knot the ropes so they won't choke him."

I climbed down the ladder from the loft and that old bull just hung there as I reached towards his nose. His eyes were watching me, his breath wheezing in and out, creating a fine spray of mucous on my hands. I felt the humbug snap in place and felt its spring take hold. I loosened one of the ropes a bit and began making a knot.

At that instant that bull came to life! He took one jump back, then thrust forward, right at me. It seemed in slow motion, I saw the humbug crack, split apart, pulling free of his nose! In total panic I tried to jump back up the ladder, but slipped instead and fell right down into the hay manger, near the bull's feet. I tried to burrow low into the hay to hide. My nose filled with the smell of his sweat and fury and I could feel my own cold fear climb up my backbone. He hit and splintered the logs on the side of the manger, while I scrunched even lower. He must have noticed my movement. He backed up again, ready to charge me, but giving me enough room to make another jump for the ladder. This time I made it. As I scrambled out of his reach, I felt the full force of the impact as his head hit the outside wall, cracking it just below my frantic feet. I don't remember what Bob was doing or how I got to the top of the ladder. I was too damn scared! All I know is, I must have been moving very fast.

Somehow, Bob and I got our ropes tight on his neck again, but that old bull just wouldn't quit. He kept jumping, trying his best to charge us. We said, "Okay, old man. No more fun," and we both pulled with everything we had, choking him right down until he was almost out. When we let him down, he knew he was beat, yet he tried one final resistance manoeuvre.

This fourteen-hundred-pound, exhausted beast sulked. He lay there on that barn floor as limp and heavy as massive stone. He watched us, but he didn't make a move as Bob and I walked around and around him, figuring our next move. I think he knew he was making almost as much trouble for us this way as he had been when he was fighting us.

We put a heavy logging chain around his neck, hitched another chain around him, then attached both to the team of horses, who had all they could do to pull him onto the sleigh. We made sure we tied him down good with the neck ropes because, even though he was quiet now, we didn't trust him one bit. We had a rifle with us and, at this point, if he'd given us any trouble we'd have shot him.

We arrived home that night when it was just beginning to get dark, securing him behind our barn, tying his neck ropes to two trees. The next morning he remembered we meant business, so we didn't have too much trouble putting him into the barn where we kept him well-secured in his stall until spring. That's when Bob and I built a new corral for him, locating it about five hundred yards from the house, across the road. As we acquired cows, we put them into his corral to be bred, then sent our dog to bring the cows back to another pasture. That way, we didn't ever have to go near that beast of a bull again, even though we kept him for three years.

Finally, he gave me one more scare, this time serious enough to mean the end of him. When Bobby and Andy were about five and three years old, I saw them walking, innocently picking flowers inside the bull's corral. We had warned them not to go near the bull—of course—and they weren't near him. But they were inside the corral, and I knew how fast that bull could move. The children had no idea the danger they were in. I felt that same cold fear I'd felt before, when we captured that bull. I was relieved when the children came running to me as soon as I called them, but this incident finished it for me. I went into the house to Bob and Mary and told them what had happened.

"Look," I said. "That bull's going to the market. We're not going to have the kids in danger like that!"

They both felt as I did. The next day, we killed the bull and shipped him to market, even though we knew that, due to the depression, we would get very little money for him. I felt only relief when we shot him, not one regret. I knew he'd had a good life while he was with us, in spite of his miserable nature.

By now, we couldn't get a decent price for anything we wanted to sell. This was the deep depression, with most people having to rely on welfare. We couldn't even sell a chicken, we just had to live on what we grew. That's why I don't agree with people today who say eggs are bad for you. During the depression we went for long periods with each of us eating five or six eggs a day, because eggs were all we had. It didn't pay to kill a steer because twelve-hundred-pound steers sold for almost nothing in those days—that's if you could find anybody with money to offer for it.

We milked twenty cows, shipping the cream on ice to market, for pennies. We fed skim milk to our sows, fattening them up to sell, knowing we'd be getting only seven or eight dollars for a whole pig, and that, only if we shipped them to Armstrong or Kamloops. You couldn't sell anything in Salmon Arm. We traded, mostly. I'd help cut a neighbour's grain and he would trade me a calf. Bob got work with the government, building roads, but even the government didn't pay money. For those that owned property, the government'd give credit towards taxes. They'd also give some clothes, or vouchers for food.

Jobs were so scarce, after a while it seemed only my brother, myself, and one other neighbour weren't living entirely off the government. It's not nice to live in a community like that. People changed. Some people got crooked, some got lazy and wouldn't lift a finger to help theirselves. But we hung on. We accumulated a lot of livestock and took comfort from that. We knew having cattle would some day be like having money in the bank. We had sugar plum trees, Italian prune plums, and some apple trees—Mary preserved all the fruit. We shot pheasant, deer, and grouse. Because we all worked so hard, my brother's children never wanted for nothing. Unlike many at that time, we had hopes of better days ahead.

Then, the following spring after getting some of our crop in, it was time to do some more stump clearing. We used the same method we had used successfully, before. This time I placed only three small sticks of dynamite, plus one large stick in the holes Bob had prepared. I lit the large stick, then turned around and—to my surprise and horror—saw Bob had touched a match to one of the small sticks.

I said, "Get the hell out of there! There's only a six-inch fuse on the big one!"

Bob didn't respond . . . maybe didn't hear me. On impulse, I grabbed him and carried him, desperately. (Christ I was strong in them days.) I made it to about a hundred yards from the dynamite. As the roar of the blast blanketed us, I threw Bob down, then threw myself on top of him. The clay, brush and rocks landed all around us. In the silence that followed, I felt my body and realized I didn't get hit with any rocks or sharp pieces of wood.

Then I heard Bob groan. Somehow in the scramble and fall, he broke his leg down near the ankle. It was a bad break. He was in a great deal of pain—both on the way home and later in the hospital in Salmon Arm.

Two days after I'd taken him to hospital, his wife Mary and their little son visited, finding him still in agony. His face revealed the extent of his pain when I walked into his hospital room. I was concerned and also confused, because I knew pain usually settles down after a broken bone is set. I lifted his blanket and saw immediately what was wrong. His leg had been placed to rest on top of a box, with his heel bearing most of the weight. This position was forcing the weight of his leg to put pressure on the break, above his ankle. I stormed out of the room and found a nurse, bringing her back to view his situation.

"Christ," I said, "A vet would know better than do that to an animal!" I was really disgusted and angry with the stupidity.

Instead of helping him, she muttered something about being right back, so I found a cushion myself and set him up comfortable. During our visit, the agony slowly left his face.

Later, as I was leaving, Bob smiled saying, "Christ, I'm glad you came in."

During my other visits, Bob and I talked of our satisfaction with our successes on the farm and our hopes of a continued peaceful existence. However, those hopes were soon dashed by McArthur.

I suppose McArthur began to figure out that he'd given us a very good deal. He must have worried about losing control of his place, wanting a bigger profit. Then, he began trying to back out of our agreement.

He began billing us for groceries purchased in his store, instead of deducting them from the $25.00 credit he promised us. Bob spoke to him about it, but McArthur wouldn't listen, claiming he'd never promised us any money for groceries or anything else. I told Bob not to worry about it.

The next day I went into town to see McArthur, meeting him on the stairs to his office, above the store. He knew he was in the wrong, because before I spoke he went red in the face and started flailing his arms around shouting, "It's all off! It's all off! We'll have to start again!"

I said calmly, "Mr. McArthur, you'd best go in your office and sit down before you have a heart attack." I followed him back inside, then I asked, "What's all off?"

He said, "Our deal's off. We'll have to renegotiate."

Then I told him I'd kept every one of his letters and consulted a lawyer who had told me those letters were as good as a legal agreement. I told him we would move out if he wanted us to go, but he would be legally bound to pay us for our moving costs, for all the work we had done, for the horses we broke, and still pay for the groceries and the coal he had promised to provide.

I said, "You're worrying me even less than you worried my brother with your threats. I'm single, free to move and go where I want . . . but mind . . . before I go, you'll pay us what you owe us."

Then he started to calm down, saying it would be okay. He decided to acknowledge our agreement, but only because he knew we'd be damn sure he didn't get away with breaking it. Later, we discovered that McArthur was much like our old beast of a bull. He'd calmed down but he hadn't given up.

His next strategy was to arrive at the farm with his son holding a branding iron in his hand. I climbed off the horse I was riding and he started talking, with an attempt at a smile on his face.

He introduced me to his son, Gordon. The old man said to me—sweet as pie—"Dave, you haven't got your cattle branded. What about if we brand them for you?"

Smiling also, his son Gordon said, "This is the brandin' iron, Dave. Isn't it a nice one?"

I looked at the iron and realized it was the brand of the oil company the old man had told me his son worked for, down in California. It was a seven and six combined.

I said, "I'm not using that branding iron. These are our cattle."

With that, Gordon got nasty, calling us thieves. Next, he tried to assert his rights saying, "You have to do what I ask. I'm the owner of this place!"

I said, "I don't know nothing about you at all! You were never mentioned in our deal with your father, or in your father's letters. Your father never even told us he had a son!" With that, I climbed back on my horse and moved away, leaving them standing there.

The following week, the old man came out again, this time with the proposition that we teach Gordon how to farm, by letting him work with us.

I told him, "No thanks. We'll have none of that. We'll just work on our own."

Soon Gordon McArthur, accompanied by his Scottish wife, came out looking for a place to build a house, on his father's land. They chose a sight down near the road, but away from us. It took them about six months to finish the house, and during this time Mary became friendly with Gordon's wife.

Mary hadn't had a close woman friend for some time, so you couldn't blame her for being pleased about their coming. Living out of the town, she only occasionally visited with other women, sometimes at the dances held upstairs in the agricultural hall. What with caring for her young family, feeding us three square meals a day (when butter was churned by hand, and "preparing" a chicken for dinner included killing, plucking the feathers, and drawing the insides from the bird), she didn't have much time for social gatherings. Even though women worked so hard, it amazed me that many farm wives belonged to the Womens' Institute, which—among other things—arranged a display of baking and needlework and crafts, once a year in Tappen. Only Bob and I were uneasy with the coming of McArthur's son's family, because we didn't feel we could trust the McArthur men.

Until Bob broke his leg, we enjoyed our life in Tappen. It was a free life. Bob and I were our own bosses, lucky to have the means to keep us off welfare, thanks to our agreement with McArthur. But that broken leg took a long time for recovery. The winter after, Bob still was not back to working his full share of the load. Mary helped me with the harvest, as best she could, but still I had to work long into the evening every night. Sometimes I'd fill forty sacks of grain after milking the cows. Then I'd load those sacks, ready for hauling to town the next day. These were long days of hard work, and I was getting worn down from it.

After Bob got well, we continued together for about another year, until the thrashing was done. Finally, I said, "Hell, I can't stand this set-up anymore. I'm tired of having to continually out-wit McArthur. I'm going to the bush near Monte Lake, logging for the winter." Which I did. There were no hard feelings when I left because Bob and Mary had known I'd be leaving some day.

While I was logging, Bob had a funny, but exasperating experience at an auction. I left him with the responsibility of selling a five-year-old mare and her daughter, a well-matched sorrel team, with beautiful silver manes. I'd acquired the horses by trading. First Bob broke them for me, then he and a neighbour took them to a farm auction in Salmon Arm, pulling them behind his wagon. At the auction, he and the neighbour planned to run the price up, by having the neighbour bid on my horses. The problem was, they didn't arrange when the neighbour should quit, and he kept on bidding until all the other buyers backed off. Embarrassed, they bought my horses for themselves and meekly led them home again, behind their wagon.

This experience is the reason Bob asked me to do the dealing with a farmer—a potential buyer whose farming ability I respected—who knew good horseflesh when he saw it. Bob had done a wonderful job of caring for my silver-maned team. In fact, they looked so good I decided to ask $300 for them. The farmer tried to bring me down, but I stuck with my price and he eventually bought them, giving me cash, right there. He knew he wasn't cheated. He knew the value of the animals and, because of our reputation, he trusted us.

After I left, it wasn't long until Bob announced that he too was through with McArthur's place. Because he was a canny Scot, he had accumulated enough money—in spite of the depression—to rent another farm on the main road to Kamloops, down below Tappen. He moved Mary, his family, some livestock, and they lived there happily for some time.

When the logging shut down and I wasn't working, I survived by using my skill at bargaining to earn my keep. I'd travel and buy up cattle, pigs, or anything I thought I might resell, and put them on Bob's farm. Whenever I visited his new farm, I helped him with the work, though my roving nature eventually lead me to move too far for this to happen often.

CHAPTER SEVEN

Kamloops, BC

Is Fortune's fickle Luna waning?
E'en let her gang!
Beneath what light she has remaining,
Let's sing our sang . . .
      —Robert Burns, "To James Smith"

In 1933, I was lucky to find a summer job I really enjoyed, and I stayed with it for about three seasons. The British Columbia government and the Farmers' Institute hired me to provide stallion services to mares, along a specific route through the rugged sagebrush terrain of the Kamloops area. The Farmers' Institute was a political lobby group, as well as a means to educate farmers about new technology and improved methods of farming.[1] I rode my own light bay cow-pony Dobbin, accompanying a three-year-old dark bay stallion, a draught horse. There hadn't been a stallion around the Kamloops area for twelve years, so I'd hit on a good thing. Many farmers had mares needing the stallion's services.

Mr. George Jackson was the farmer who hired me, assigned me the stallion, and reported our earnings—in stud fees—to the Institute. He paid me a wage and it was up to me to pay the farmers en route for my room and board. On the weekends of my first summer on the job, Jackson arranged for me to stay at a farm owned by Senator Bostock, where our stallion bred fourteen of his mares. Regularly, on the same day every week, I would visit and stay overnight on the other farms of my route, until we were satisfied the mares were in foal. My customers made me welcome and I enjoyed their sociability.

My first summer, I was expected to breed fifty mares and have a 60 percent pregnancy rate, which would entitle the Institute to a grant of three hundred dollars. Without broadcasting what I was doing, I bred seventy mares, then picked out the ones that were pregnant, making it appear I'd had a hundred percent pregnancy rate. I chuckled to myself with amazement at my success. Everyone was convinced I was a champion horse-breeder.

As a result, at the end of the season, I took four hundred dollars cash (in service fees) home to Mrs. Jackson and handed the bills to her.

"Four hundred dollars!" she laughed. "I've never seen the likes of this before! Did you rob a bank?" She was thrilled because times were so hard.

The members of the Farmers' Institute got their full three-hundred-dollar grant from the government, so they were happy. Old Jackson gave me a few extra dollars with my pay, so I profited too. At first, I thought Mrs. Jackson must have told him to give it to me, because I had the impression he wasn't the kind of man to part with a dollar he didn't owe. He was a big square man with wedge shoulders, a slow talker who seemed to carefully select each word he spoke, in a voice that sounded as if it came up from his feet. Yet, when he got mad, he was a wild man. After he got to know me, he helped finance one of my deals by borrowing one hundred dollars from the bank—gave it to me with no questions asked.

I used his money to buy eight young sows (already bred), twenty pigs (three months old), and a two-year-old mare—all offered by a farmer who was desperate.

"One hundred dollars for the lot," he said, but I didn't have the hundred dollars. This was a most impressive deal, because at this time I was buying cows for fifteen dollars apiece and selling them for just a little profit.

I happened to meet George Jackson on the street just after I left this farmer. I told Jackson about the offer of the livestock, not revealing that one hundred dollars was the full price. I told him I needed a hundred dollars to close the deal, and he just nodded, heading straight for the bank to arrange the loan.

The next day, Mr. Jackson asked me again about the animals. When I told him they were top quality, he said he'd been looking to buy four sows, so he'd buy mine. (By this time, he knew my dad had taught me well and he trusted that I knew good livestock.) The deal became even more satisfying when Jackson paid my price—one hundred dollars for just the four sows. I felt good because the sows were worth the money, plus we were now square for his loan.

I took the other four young sows, the twenty pigs, and the mare back to Bob's farm. He broke the mare and sold it as quickly as he could. He fattened the pigs with surplus skim milk from his cows, then he sold them. He paid me my money back and a little more. I didn't ask how much he got for the animals because he had a family to bring up, and his wife Mary had treated me well.

Like other women of that time, Mary worked hard right along with her man. George Jackson's wife was another good example. Hetty was a kindly soul with an enjoyable sense of humour that was much-needed to survive the farm work and the raising of eight children. She was also head of the Womens' Institute and a very smart woman.

But getting back to my first summer as a horse breeder, I must give credit where credit is due. It was "Ramsey McDonald" that actually did the breeding. My assigned stallion had been registered with that name, which caused no end of fun with the farmers on my route. It was my job to keep Ramsey well-fed and healthy, to make sure the mares were ready for him, and to get him to all the right places at the right times. I worked hard to give that horse the best of conditions; then he had all the fun with the breeding. I'm sure Ramsey enjoyed his job even more than I enjoyed mine.

We were so successful that first summer, that Mr. Jackson must have thought I needed more of a challenge. At any rate, he sure surprised me because during the off season he sent Ramsey back to his home in Leduc, Alberta, and assigned me a new six-year-old stallion named Ben.

This second summer, my weekends were spent on "The 101" ranch, and my route included farms nearer Stump Lake. Travelling here, it didn't take me long to realize that Ben was not nearly the stallion that Ramsey McDonald had been.

Oh Jesus! I thought, *Am I having troubles!* Almost every mare that was bred had to be repeated. Christ! I was almost grey-haired by the time I finished the first go 'round. Then, one day I met a fellow who recognized Ben and explained his condition. This fella had known the farmer who owned Ben, in Alberta.

"He's such a good-looking horse," the man said. "Too bad he didn't leave any colts."

"What?" I said, "What de ye mean, no colts?"

"My Alberta friend gave Jackson a good deal for the use of Ben, because he didn't want the horse back. My friend didn't like Ben and I think he mistreated him. That could be why Ben doesn't breed well."

I thought, *Damn! That Jackson never told me this! And him a Scotsman, too!*

I decided food was the key. The only thing to do was build Ben's stamina and success by feeding him royally. *If this doesn't work, I'm sunk,* I thought.

On weekends—because the farmer I stayed with had dairy cows—I arranged to feed skim milk to Ben, and I also gave him eggs. I put him in a special sunny paddock, in front of the house. Ben seemed surprised at the change, but loved the milk and his special treatment. He drank eight gallons a day, sixteen gallons on a weekend, night and day. This wasn't expensive then, because dairy farmers often threw away surplus skim milk, after they'd skimmed the milk for cream or making butter. Any cracked or questionable eggs were tossed, too. So Ben got them.

It wasn't long before he responded. The first horse he made pregnant was my own saddle pony, Dobbin, and from then on everything was "Jake-a-loo." I stayed an extra two weeks at summer's end, and got a full, successful season out of Ben. Ben was also pleased with himself, I could tell.

Throughout the summer, Jackson must have been anxious, because he kept writing me requesting updates on Ben's progress.

For spite, I didn't answer his letters and wouldn't tell him which mares were now in foal. He had to ride the route asking the farmers. At the end of the season, I drove his car while he collected the farmers' fees. I grinned as he discovered what I already knew. Ben and I had 78 percent of the mares in foal.

When next I returned to Jackson's house, I said hello only to his wife. I was still peeved with his trading of Ramsey for Ben. Jackson looked at me sideways, all through lunch.

Finally I said, "Why didn't you tell me about Ben's problem? If you'd told me, I could have fixed him sooner." He didn't answer. He just looked uncomfortable.

Later, Hetty Jackson told me that Jackson had controlled his temper because he didn't want me to quit. She said, "George says you're the only breeder that makes the Institute any money."

By the time the third summer season came around, I really felt the farmers we serviced were my friends. You see, I didn't just breed the horses, then sit down on my behind. I used my farm training and experience to help the families I met. This is the way my father behaved in Scotland, and this is the way I did it, too. Often, I helped with the milking and even helped the wives with the dishes, if that's what was needed. I enjoyed helping them.

One family, the Deleeuws,[2] were good farmers, originally from Belgium. Their two twins, Hilma and Hilbert, and their other son, Andre, took me to their pig sty. The children were worried they were going to lose their litter of newborn piglets.

"We don't know why our sow won't feed them," they said.

I knew what was wrong when I watched the sow reject the first piglet offered, but I didn't explain. I just told the children to help me gather the piglets and take them to the house, "so the sow won't hear them squealing." The kids looked unhappily at me as we walked back to the house.

"What are you going to do with them?" they asked.

"Save their lives," I replied.

I got a big pair of cutting pliers and, one by one, nipped the ends off the piglets' sharp little teeth. The sow was rejecting them because their teeth were hurting her. Though they squealed bloody murder when I cut them, if the teeth had stayed sharp, the piglets would have died.

Mr. Deleeuw and his wife accompanied us when we returned the litter to the sow. I held one of the piglets to a teat. At first, I felt the sow stiffen as if to reject it, but when she felt it suckle, the sow relaxed. It felt good to see the whole family so happy, smiling with their arms round each other as we watched the hungry piglets eat their fill.

The Deleeuws, like many families on my route, really appreciated any help that was given. I enjoyed ambling through the dry Kamloops hills on horseback, meeting so many kindly people. I continued buying horses for myself, whenever I could afford them. I boarded some of them on these Kamloops ranches, until I wanted to retrieve them for resale. Working and getting to know people in this area led my life in new directions.

I became interested in breeding racehorses—but never in betting on them. I bought a seventeen-year-old racehorse in Kamloops from a fellow named Hotchkiss. I bred it to a thoroughbred stallion named Dolon, owned by a wealthy man named AustIn Taylor—very well-respected in horse-breeding circles. This breeding cost me twenty dollars then, but today it might cost as much as $17,000 for the same quality of stallion. Dolon bred many offspring, some of them good racers and some of them hard to handle, because he was a high-strung horse, cranky. He'd as soon bite or kick you as look at you. I was glad I wasn't his handler.

Jackson started me going to the annual bull sale at Salmon Arm, an interest that continues to this day. This bull sale has been happening every year since 1918 and includes not only stock, but also offers produce, farm implements, and harnesses for sale.[3] Jackson urged me to go with him because he needed a driver for his car. He had never learned to drive. The trips to the sale became a habit and I still return there most years, right up to the present. The big difference between my going then and now, is that now I'm driving alone, in my own car. The bull sale still interests me, but it's a sad time, too. Every year I see fewer of my old-time farmer friends, but that's a natural change. I have to expect that.

In August about 1936, travelling with Jackson led me to another farmers' gathering, this one at the Pacific National Exhibition (it was usually called The PNE) in Vancouver—another interest that gave me pleasure for many years. After retirement, I spent several years as Superintendent of Livestock for the fair. Of course, I didn't know any of this when Jackson and I left Tappen.

In 1936, Jackson and I were travelling in a Model-A Ford, accompanied by an Irishman named Mike Lynch, and a man named Patterson, who did most of the driving. Patterson and I sat up front, Lynch and Jackson were in back. The animals were being shipped by train to Vancouver—our three stallions, a six-horse team, and eight show horses. We planned to stay ten days in Vancouver and then go to the Victoria Fair.

The trip from Kamloops to Vancouver took two full days, travelling thirty miles an hour on roads that, at times, made your hair stand on end. Long strips of washboard road made the teeth rattle in our heads. In the Thompson and Fraser River canyons, sometimes we were looking two hundred feet straight down. Sometimes we were riding on a road so narrow the car couldn't even turn around. If we'd met a car coming the other way, I don't know what we'd have done. Back up, I guess.

At the end of the first night, we stopped in Boston Bar, staying two to a cabin in the same order we'd been sitting in the car: George Jackson with Mike Lynch, me with Patterson. At about 2:00 A.M. Patterson woke up and shook me awake.

"Listen!" he said. I heard a fearful snorting sound . . . very loud. "It must be pigs, big pigs," he said. "Them's the only things I know makes a sound that loud. They must be awful close."

With that, he went outside to find them. Moments later, he returned just as I was getting my boots on to follow him. "My God!" he said. "That's not pigs, at all! It's coming from the other cabin! It must be one of them!"

The next morning, Mike Lynch was wide-eyed and full of energy. Jackson was bleary-eyed. It wasn't hard to figure—Mike was doing the snoring. He had a snore so loud, you could hear it miles away. I'd never heard anything like it in my life!

When we hit Vancouver, we discovered Patterson didn't know how to drive in the city. When he was afraid in traffic, he'd tense up. When he tensed, he'd press his foot down on the gas pedal. We travelled faster on city roads than we ever had in the country, just like one of those Keystone Cops chases. None of us ordered him out because we'd never driven in city traffic, either, but we were all very relieved when we arrived at the gate to the PNE grounds. Patterson may have been more relieved than any of us.

We found our shipped horses had arrived safely, ready for their successful PNE debut. The three stallions, our six-horse team, and eight show horses performed very well and made us proud. They won several medals and received many compliments before they climbed into the trailers for the ferry trip to Victoria.

When we left Vancouver, I drove. I did a pretty good job because I'd practised a bit during our short stay. Nobody wanted a repeat of Patterson's driving on city streets. As we wound our way through Vancouver streets towards the Victoria ferry dock, I had no way of knowing this trip would eventually lead me to Colony Farm, one of the largest government farms in British Columbia. Within days of arriving in Victoria, Pete Moore, Colony Farm's superintendent, would approach me and offer me a job—a month's work breaking horses in Coquitlam.

My first impression of Pete noted only that he had a dark complexion. He stood about five foot ten, with thick black hair, his chin darkened by a five o'clock shadow. (Later, I noticed this shadow would reappear only a short time after his morning shave.) Eventually, when I saw him angry, I saw he had a side to him darker than his hair: his black Irish temper. I heard he had once been a heavy drinker, but by the time I met him, this was no longer a problem. I also discovered he was quite a ladies' man.

In Victoria, neither of us knew much about each other, but he liked my work and I liked his direct manner. Eventually, I learned he'd come to BC from Truro, Nova Scotia, where he'd been born.[4] He'd gone to the Truro Agricultural College, then got his bachelor of science in agriculture from Toronto University. He'd become superintendent of Colony Farm in 1914. Though he was well-schooled, he was also very practical. Like me, he loved working with Clydesdales and dogs. He was interested in giving the livestock the best of care and good breeding, so they could become the best stock possible. He knew what he was doing, and I respected that.

I was never afraid of Pete Moore. He was always straight with me and I always told him what I was thinking, even if it wasn't always what he wanted to hear. Later, as we worked together, we gained mutual respect. I learned he wasn't liked by many who tried to pressure him to get what they wanted. He didn't give in to their tactics. They hated him, but I liked him for that. What kind of a man is it who buys his friends with favours? Though I didn't know it then, when I accepted Pete Moore's job offer, I was also accepting a working friendship that would support me through difficult times for years to come.

At the end of the Victoria Fair, I went back to Tappen and told Bob I'd be taking a month off from helping him. I was going to Coquitlam to work for Pete Moore at Colony Farm.

The one-month job extended to three months. Pete asked me to continue as a farmhand, looking after the work horses and running the heavy machinery needed for planting. For me, the three months came to an end far too fast. It was almost Christmas when I went to see Pete Moore to say goodbye and thank-you.

Pete responded, "What do you mean? What's the matter, don't you like it here? We're hoping you'll decide to stay. Why don't you go home, settle your affairs, and come back here to work permanently?"

He had a big smile on his face, the first time I'd seen his smile so wide. When I heard his words, I could feel a smile growing inside me, too. Over 1,400,000 Canadians were out of work and starving, but I had just been offered a permanent job.[5] Finally, after the long years of living by my wits, Pete Moore was offering me a job with a future, doing what I loved. I couldn't have been happier.

During Christmas with Bob and Mary, they both encouraged me to accept the offer. So in gratitude, I rounded up the horses I'd bought at the various farms where I worked as a horse breeder and gave them to Bob to sell. I told him to keep all the money for his family. I knew I wouldn't need it, now that I had a job that paid well.

I felt optimistic and truly blessed.

CHAPTER EIGHT

Colony Farm

Before ye give poor Frailty names
Suppose a change o' cases . . .
  —Robert Burns,
      "Address to the Unco Guid"

I started working at Coquitlam Colony Farm one month before Christmas, 1936, expecting this job to have much in common with my job at Macdonald College Farm, in Quebec. Though there were obvious similarities—the natural beauty of both settings, the raising of animals, the cultivation of fields—life and work here was different. More complex, more challenging.

By the time I arrived, Colony Farm produced food not only for the huge provincial asylum (Essondale) but also for the large provincial tuberculosis hospital in Kamloops (Tranquille), for Woodlands School for the Retarded (in New Westminster), and for the large provincial prison (Oakalla). Here, instead of young students, Colony Farm's trainees were patients—troubled patients ("inmates" they were called)—of a provincial mental hospital.

The use of patient (or inmate) labour was nothing new, either at Colony Farm or in British Columbia. Patient labour had been used at least since the first Provincial Asylum for the Insane was established at New Westminster in 1878, long before I arrived on the scene.[1] Able-bodied patients had been required to provide volunteer labour for a four-acre garden area; they had helped clear the thousand acres that became Colony Farm; and patient labour also helped build the dikes and establish the drainage to reclaim the farm's riverside marshes. Then—after the old asylum was moved to its new site adjacent to Colony Farm—supervised patient labourers tilled the rich, fertile soil for planting and harvesting.

Notes to chapter 8 are on page 240–41.

BC's first Residential Medical Superintendent, Dr. R.I. Bentley, supported their involvement.[2] The guiding beliefs of the time were "the idle confinement of imbeciles must be rather provocative of further evil than of cure," and "Treat a man as a responsible citizen and he will endeavour to justify that confidence."[3] Bentley and his 1905 successor, Dr. C.E. Doherty, argued that work on the land would not only improve the physical condition of the patients, but also help cure their insanity by making them feel useful. Furthermore, they believed the newly acquired farm skills would fit former patients for immediate employment when discharged.

Dr. Bentley sold his ideas to the government of the day by stressing cost savings on food and the fact that patient's nutrition—improved by fresh farm products—would help them recover and be ready for discharge faster. And almost as important—since 1913 when Colony Farm surprised everyone by winning all but one of the fourteen livestock championships at the Ottawa Fair—the patient labourers were learning methods that produced championship stock. In this way patient labourers helped to make Colony Farm a source of pride and improvement to British Columbia's agricultural development.[4] Though Dr. Doherty had been gone for twelve years before I arrived in 1936, he was still talked about, remembered as a very congenial, good man, very modern in his thinking.

As far as I was concerned, I agreed with the ideas both previous superintendents shared about the use of patient labour. Their ideas made sense to me when I first heard them, and they still seem common sense to me now. My father would have also agreed with them.

He had told me long before in Scotland, "Davy, mind you don't keep horses penned too long indoors because it ruins them, teaches them bad habits that are hard to break. People are no different, in this respect."

I believe it harms people to pen them in hospital rooms—or even offices—day after day. Even healthy people will wither away. Fresh air and exercise are needed as much by humans as they are by animals.

When I worked the fields of Colony Farm, there were seven work gangs of patients, with ten people in each gang. Because these people weren't mentally strong, it took almost as many attendants to supervise them as there were patients working. Twelve patients worked in the cow barns, with the cowmen supervising them as they cleaned the barns and brushed some five hundred milk cows, twice a day. Of the five hundred cows, 230 of them would be in milking condition, the rest would be young ones not yet mature enough to milk, or not giving milk for various reasons. Each milking dairyman milked the same fifteen cows, three times a day, a big job. Two patients worked in the calf barns, feeding calves and cleaning the stalls. Six patients worked in the piggery, cleaning the building that was home to some fifteen hundred pigs.

Patients also worked in the fields, where I worked for the first three years I was there. These fields were built by nature over thousands of years, from the silt of two great rivers, the Coquitlam and the Fraser. Some of the soil we were tilling was probably carried downstream by the Fraser, all the way from Prince George. I spent my working hours mostly ploughing seventy acres of potatoes, drilling twenty acres of mangles (turnips used for cow feed) and looking after the Colony Farm Clydesdales, horses I most enjoyed. Only those patients designated as "on parole" were allowed to work by themselves, and two of them worked in the horse barn, cleaning it under the supervision of the stable man. I would see and talk to them when I brought the horses out or in.

During these years, the patients I saw just loved the farm work. I remember they would gather in a group at the doorway of their building, waiting for their supervisors to come and take them to the jobs. Some of them were very good workers when they were properly monitored. They were well-cared for, working about six hours a day. If it rained, a steam whistle was blown to signal the patient work crews to come inside. If it rained for a few days in a row, the patients would cause a great uproar in their quarters. Locked inside, they became wild because they couldn't stand being cooped up. Knowing what I knew about how much they loved the work, it made me all the sadder to see the work privilege denied them, a few years down the road. But more about that later.

About three years after I arrived, a hard-working truck driver had a heart attack and I replaced him. At first, I expected to be out of the fields only until he got better, but he died, poor fella. So for the better part of the next year and a half, I was a truck driver hauling heavy loads in a three-ton Nash Diamond T. That truck had to back up to turn around, and had a motor that raced if I drove it thirty-five miles an hour. It was a good, sturdy farm truck, though it had no markings on it to identify it as a government truck. It was so sturdy, it was still in use for several years after I retired.

This is the truck I drove while working with Bill Budlang.

When I drove this truck, I had a patient helper, Bill Budlang, and I had a lot of fun with him. At first, he seemed very strange to me, but I learned to care for him. Bill was over six feet tall, strong, all muscle. I don't know what was the matter with him, but he spoke very few words. Over and over again, especially when he was in a new situation, Bill would make a fist with his right hand, hit his neck, and grunt loudly. He also kicked at his own legs quite a bit, even though he wore big boots with steel heels. With his kicking, one day I noticed he was wearing a hole in his leg, making it bleed.

"Cut that out, Bill!" I said. "Can't you see you're hurting yourself?" But Bill just grunted louder and kicked all the more. I felt sick about him injuring himself, but no matter what I said, I couldn't make him stop.

I decided to try to help him. First, I went to his ward and asked his nursing attendants if he could be given some softer shoes, but they said they could do nothing. Frustrated, I spoke to the doctor in charge of patient workers, whom I knew to be a caring man who frequently checked to see how the patients were doing. He gave me a permit to go to the store and buy Bill some shoes with inch thick crepe soles, so he couldn't hurt himself anymore. You should have seen the pleasure on his face the minute he put those shoes on. He walked around in circles for a bit, stepping like a stiff-walking heron, listening to the squishy noise he made as he trod on the pavement. He was enjoying the shoes so much he didn't want to get into the Diamond T and go to work.

The first time we were in the truck together, I saw the joy Bill found in his volunteer labour. It was in late August, getting on to harvest time, when I drove to Alexander Street in Vancouver, going to the American Can company for a large load of gallon cans (needed for preserving the food grown on Colony Farm). On arrival, I found out American Can stored cans sixty-four to a carton, with the cartons piled four high, too high for me to reach them for loading. So I brought Bill over and showed him the problem. Without a word he picked up the top carton, holding it high in the air. Then as if he were celebrating his success, he began pushing the carton up and down, grunting and dancing along with it, causing me to dance along behind him holding my arms up to catch the carton, if he should drop it. I felt foolish even before I noticed a group of men from American Can were watching us and laughing like hell. My face flushed with embarrassment, but I noticed Bill was in his own world. The laughter didn't bother him one bit.

On our return trip to Colony Farm, I swung in a wide turn off Alexander Street onto Powell Street. Then the Nash Diamond T had to back up, jockeying to make the turn. When I leaned out to look back and signal anyone behind me, another driver started shouting loud and furious, "Get that damn thing out of here, you dumb farmer!"

I saw red. Bill watched, hitting his neck and grunting, as I climbed out of our truck shouting and shaking my fist. I yelled back, "Climb down here and I'll show you what a dumb farmer can do!" The guy's face got redder, but he stayed in his truck. I thought I saw a trace of a smile on Bill's face as I climbed back into the cab of the Diamond T. From then on, Bill and I got along just fine.

One hot day, I stopped for the first time at a soda fountain on Douglas Road in Burnaby, for my favourite lemon milkshake. I left Bill in the truck, because of the effect his behaviour had on those who didn't know him. As well as my milkshake, I ordered a double cone for my partner. When the waitress handed me my milkshake, she asked who was getting the cone.

"My partner, out in the truck," I said. "I'll take it out to him."

"Don't trouble yerself," she said. "I'll give to him."

Without thinking, I thanked her and watched her walk outside. Then I went back to the enjoyment of my milkshake. Within seconds, she was back behind the counter breathing hard, with her face white as the first snow.

"My god!" she said. "What kind of a man is that?"

I'd forgotten how Bill always moved with a jerk and very quickly. She said she had just reached up to pass him the cone through the window, when he started his grunting. With the look on his face, his noises, and his grabbing the cone so fast, she thought he was going to take her arm off.

Still puffing she said, "I've never moved so fast in my life!" I apologized and, though the lemon milkshake and double cone became part of our summer routine, from then on I took Bill's cones to him myself.

Another time I looked out the window of the ice-cream parlour and saw a burly motorcycle policeman and his partner in his sidecar, both stop and walk around our truck, looking for the identification signs on the doors. As I stood to walk out to the policemen, I told the waitress and another guy in the parlour to watch the fun. Before I had moved very far, one of them opened the door on Bill's side of the front seat, and my helper did his usual, grunt, grunt, grunt.

Christ, the policeman jumped back about three feet in surprise, knocking his partner on his ass. When I walked over, both were brushing the dust off themselves, saying to me "We're going to arrest that fellow in there."

I looked at the six-foot policeman and his partner and said quietly, "Though ye're both big, I'm not sure ye're men enough. Ye saw how big he is, and I'm warning ye, he's a patient at Essondale."

"Are you in charge of him, then?"

"Aye," I said, "but it's taken some time for him to get to know me."

With that they looked at one another, then turned away saying, "Let's get out of here!" The sidecar passenger jumped back in his seat, as his partner kick-started their motorcycle. They were off in a cloud of dust as I returned to finish my milkshake, accompanied by the laughter of the waitress and the other customer.

Another time, the old Diamond T and its lack of marking provided more laughter. It was in the early morning when Bill and I were given the job of taking four Holstein cows down to Sullivan Farm for a livestock show. We had to cross the old Patullo Bridge. This bridge was so narrow, if there was a load of hay coming the other way, we'd have to wait until it finished its crossing, before we could begin.

Just as we were about to go on the bridge, two policeman waved us to a stop. One of them asked, "Where's the identification sign on your truck?"

I told him, "This truck's owned by the government, and the people who are supposed to paint the sign won't do it. I've asked more than once, so I've given up asking." He waved me on because it was a government truck, otherwise I'd have had a ticket.

I dropped the cows off, came back to the old Patullo Bridge on the way home, and was stopped again by the police for the same reason. I explained again. Then I had to go back in the afternoon to pick up the cows. The same two policemen stopped me for the third time. Tongue-in-cheek, I began the same explanation. After the first few words, the policeman's brow wrinkled and he interrupted. "Weren't you here before?" he said.

"Yeah," I replied, holding back a grin.

"Oh go to hell!" the policeman said, and I left laughing, with Bill making his usual noises beside me in the front seat.

In spite of the fact Bill Budlang spoke very few words to me, I knew he was a good man, and I looked after him. Together we hauled potatoes to Woodlands and Oakalla, part of the eight hundred tons of potatoes grown annually on Colony Farm. I worked him hard, but I always hauled bag for bag with him whenever we were together.

After I became herdsman, I heard the man who took over my truck-driving job treated Bill badly, nearly killed him. He had Bill unload a four-ton load of grain—cow feed for the milk cows—all alone. I was told this driver always left Bill to haul all the loads, while he'd go and do as he pleased. But that man got his due in a way I wouldn't wish on anyone.

Not long after he took the truck-driving job, he went blind from a disease. I wasn't the least bit sorry when I heard about his misfortune. I figured he had it coming to him for what he did to Bill. I've always believed that if I knowingly do wrong to innocents, it'll bring me bad luck.

Sometimes the job required me to drive alone, in a three ton Ford with lots of speed in it. Every Monday I had to crowd thirty-six complaining pigs into this truck. Then I'd deliver them to Swift's Canadian abattoir, to be killed and have their offals and hair removed. On Tuesdays, I'd return to collect the whole carcasses from Swift's, delivering them—according to a list I was given—to the cooks in the kitchens of the various hospital buildings. The cooks did their own butchering. This job was pretty straightforward, with little paperwork involved. Lacking paperwork, I suppose I could have been crooked, selling some of the meat on the quiet, if I'd a mind to be that way. But I wouldn't even give this a thought.

One day as I arrived back at the farm from the abattoir, a fellow who had been following me got out of his truck and returned a pig carcass he said I must have dropped. I thanked him, but as. he turned away he said, "Christ, you might have at least given me a leg!"

I said quietly, "Listen fella, if you wanted some of that pig, you should have picked it up and driven away, because I don't own that pig. It's not mine to give you. If I was back on my own farm, it would be a different story." He just looked at me and shook his head as he got in his truck and drove away.

Later I told my roommate, George McMillan (who we mostly called Geordie), what happened. He said, "If that guy had known you, he never would have asked for a payoff." Geordie knew I'd consider myself nothing, if I wasn't as honest as I'd long ago been trained to be, by my dad.

Geordie and I had known each other for a few years before he was hired at Colony Farm. In fact, he got his job partly because of me—because at this time, the hiring practices allowed me to give input. It happened this way.

When I'd been at Colony Farm just over a couple of years, I was ploughing a field when I recognized Geordie walking on the side of the road, looking very dejected. I called him over.

I said, "Don't I know you? You're George McMillan from Kelowna, aren't you?"

We chatted about the ploughing matches in Kelowna, where we'd met and liked each other. This day, I asked him, "What's troubling you, Geordie?"

He replied, "I've just been in to talk to Bob Gardner, one of the foremen at Colony Farm, about getting a job. Gardner said they couldn't use me and I'm disappointed. I really had my hopes up."

I said, "Geordie, come on back to Colony Farm for lunch with me, and we'll see what we can do."

After lunch I went to Bob Gardner myself, and told him he'd be missing a damn good man if he didn't hire George. I told him George was a friendly fellow, a good man on a farm. I told Bob I'd met George and his father, Hector, showing horses at the Armstrong Agricultural Fair. George and I had competed at ploughing matches in Kelowna. I also knew a fellow name of "French," their former boss who was an orchard man in Vernon, who held both George and his father in high regard. I knew I could vouch for George as a good farmer.

Bob Gardner was grateful for my input and he hired George on the spot. I didn't know it at the time, but three years down the road, my helping Geordie would lead to my marriage to his sister, and eventually to my having a son who became his namesake. I've never regretted helping George McMillan to get a job that day.

My speaking for him was only possible because, in those years, the hiring for Colony Farm was all done by the people who worked the farm. There was real strength in this practice because we all understood the skills and the knowledge of machinery and animals that a farm worker needed to know. We were farmers ourselves.

In Victoria, when Pete Moore saw my work and judged it good, he was free to hire me. Like my friend George, I was able to keep my job because my work proved I knew what I was doing. The hiring method at that time was a far cry from the way it was done later, when I became farm manager. By then government officials legally had to be involved in the hiring interviews, often making themselves and the process a joke.

Geordie McMillan and I continued our friendship, sharing a room upstairs in the Colony Farm arena, where prize-horse shows were held. Outside our room was a balcony, where the Essondale doctors and their friends would watch the horses jumping. About ten of us farmhands slept in this huge building and we found it quite comfortable.

Geordie was still living in his parent's home, taking his washing home on weekends, so it was awkward for him to take the bus while carrying a big bundle of dirty farm clothes. I offered to drive him, and began visiting with his family most every weekend. Of course, the fact that George's sister Peggy was living at home, encouraged me to visit. I'd been interested in seeing more of her, ever since I had first met her in Kelowna, though at first I wasn't sure if she was interested in me, at all. I told this to Geordie.

Then he said, "Dave, I'm not sure if you're going to be interested in her anymore. She's been through quite a bit since you last saw her." Embarrassed, he told me the story, explaining why they were now living in Duncan instead of Kelowna.

"It seems when Peggy was twenty-four, almost three years ago," Geordie began, "she met some married bugger while she was working at the hotel, and he made her pregnant. Being young and inexperienced, Peggy was easy prey."

Silently angry I said, "Anybody who knows Peggy would see that."

"But not mother," Geordie said sadly. "She had no sympathy for Peggy. Mom thought only of herself, I'm afraid. She was too ashamed to face the scorn she imagined would come from the Scottish Society. Mom wouldn't let up until she forced me, Dad, and all of us to leave Vernon. Though she's better about it since she saw the baby, in 1936 she as good as disowned Peggy, leaving her alone to fend for herself and her unborn baby, in Vernon. It was only after little Ethel was born, she helped Peggy by allowing her back into our house. Now, we let on Ethel is Mom's baby, so's Peggy can keep the job she has at the Duncan hotel. If the hotel knew the whole story, Peggy would lose her job."

Geordie paused and looked at me for a reaction. He was probably expecting embarrassment on my part, but he didn't get it. In spite of the fact that having a baby out of wedlock was a disgrace in those days, to me it made no difference to the way I thought about Peggy or her family.

"Hell, Geordie," I said "after being on farms around animals the way we have, I've thought these things out for myself. What's more natural than having a baby? I appreciate you telling me, but it makes no difference as far as I'm concerned. But tell me, how did Peggy make it on her own?"

"She survived because she'd made a friend, Ethel Chase, in Kamloops when Peggy worked at the hotel. In spite of the times, this friend's family had money enough to own a real estate business and a cabin, located about a mile north of Kamloops on Paul Lake. They were kindly people who hired Peggy to be resident housekeeper at their cabin. Peggy told me she felt safe during her pregnancy. She said it was a beautiful spot and I guess she enjoyed the privacy of the place. She sure admired her employer and the kindness . . . so much so she's named her baby girl Ethel, after the woman who helped her."

With the ending of Geordie's story, we both lapsed into silence as we bumped along the road to Vancouver and the ferry. *Christ,* I thought, *Annie abandoned Peggy in 1936, the same year Pete Moore hired me at Colony Farm.* With so many Canadians unemployed, how could she be sure Peggy'd find work? She could have become one of those people I'd read about in the paper, starving to death, with no aid.

I could understand how Peggy wouldn't know enough to protect herself from getting in a family way. Men in this country have access to information women can't get. We men can buy condoms at drugstores to prevent our girlfriends from getting pregnant, but neither doctors nor anybody else have the right to teach birth control to women.

In fact, I'd read in the paper, a person could get two years' imprisonment for distributing birth control information. (The common belief was that giving birth control information to women would threaten the authority of their husbands.)[5] I'd read of a young woman, a social worker back East, who had been acquitted after a four-month trial that charged her with distributing birth control information. In spite of the fact she was found innocent, after the trial this social worker received obscene telephone calls and letters, was shunned by relatives, and slapped by strangers on the street. Then I remembered, there was that bloody guy who tried to rape her to show her what it's like without birth control![6] I laughed like hell when I read she'd stopped him with a kick in the groin.

Aloud I said, "Don't worry, Geordie. We won't speak of this again."

To myself I thought, *this country's not much different with their attitudes than Scotland.* For most mixed company, all mention of sex and pregnancy's taboo, and even married women try to hide pregnancy by staying away from public places when they begin to show. How could Peggy's mother have abandoned her, pregnant in a world like this!

For the rest of the trip to Duncan, I thought about seeing Peggy again. Though I felt sorry for the hard times she'd been through since I last saw her, I couldn't help but admire the strength she'd shown overcoming her difficulties.

When we finally arrived at the McMillan home, Geordie McMillan and I must have been quite a sight. Two young fellas—him with his washing piled in the rumble seat—both of us perched proudly in my two-seater, 1929 dark blue Chevrolet Coupe. I guess it would be the equivalent of a fancy sports car today. I paid Chris Brown, of Brown Brothers car dealership, three hundred dollars for it. Chris was a friend of Peter Moore, and that told me he was trustworthy. When I bought this car, I didn't know it would be the means to the end of my life as a roving bachelor, helping me win the woman of my choice. But it wasn't just the car that led to matrimony—it took World War II to push me into married life.

CHAPTER NINE

Of Love and War

O happy is that man an' blest!
Nae wonder that it pride him!
Wha's ain dear lass, that he likes best,
Comes clinkin down beside him!

—Robert Burns, "The Holy Fair"

I'd first met Peggy when I lived in Kelowna, around 1933 when I was about twenty-nine and Peggy was twenty-three. We both attended Scottish dances in Vernon. After our first meeting, over several years I'd built a talking acquaintance with her at plough-ing matches, and enjoyed the company of her brother and father. To my eyes, she sure was a good-looking woman, though I thought we were an unlikely pair. Peggy never wanted to be a farmwife and I was never happy as anything else but a farmer. Six years had passed by the time we became reacquainted in Duncan. I was now thirty-five years old. As soon as I saw her again, I real-ized Peggy'd lost none of her appeal.

I said, "Peggy, it's good to see you again.

She replied, "It's good to see you, too Dave. And it's good you're not coming here to ask me to dance." We both laughed, remembering the first time I'd made such a request. But neither of us was laughing that first time, in Vernon, no sir.

I suppose I'd have sensed her disdain before I felt it full-blown, if only I hadn't had my usual few drinks of scotch before the dance. I drank because I was a bit shy at these affairs. A wee drop or two bolstered my courage and helped me through the nec-essary polite conversations. (Though there was "prohibition" in Vernon, you could buy liquor in the liquor store to drink "in pri-vate.") Though the whiskey helped me relax, I was enjoying the

Note to chapter 9 is on page 241.

dances more than the talk, especially the Highland Schottische, Patronella, Gay Gordon, and the Eightsome Reel.

Disaster struck as I was dancing the Edinburgh Quadrille with a girl I didn't know. The liquor had me feeling good and I gave out a whoop of joy when it was our turn to swing. I swung my partner with a real vigour to match my enthusiasm, not realizing my strength. Before I could do anything to slow myself, I felt my partner's hands slip out of mine.

Good God! I thought as she flew right across the room, hitting a post, and falling down in a faint! The whole room stopped dead as people ran over to help her. One-by-one the others—including the attractive Peggy McMillan—turned to glare at me. I quickly went over to help, offering apologies to all. To tell the truth I was in shock, thinking my dance partner was dead. Fortunately she revived and was all right, but, Christ, I felt bad about it.

It was no wonder Peggy gave me a straight-out no-nonsense refusal later, when I mustered the courage to say, "Will you dance this one with me?"

She replied bluntly, "Now, I'm crazy about dancing, but I'm not crazy about dancing with you! Thank you very much!" With that, she turned her back on me.

My pride was hurt and I was disappointed, but after what she saw me do, I can't say I blamed her, not one bit! That characteristic bluntness and honesty with her opinions was a trait Peggy never lost—and I liked that about her.

She was still that way on weekends when I visited with Geordie at his home in Duncan, and I was pleased to learn more about her. At first I had little opportunity to be alone with her and I reacted as if she was just another member of the family I happened to be visiting. But I liked the friendliness she showed me, different from our first encounter at the dance.

Then one weekend, after visiting friends in Vancouver, she stopped off at Colony Farm to see George and me, before going home. When she was ready to leave, I said, "You don't need to catch the bus back, Peggy. I'll be your chauffeur and give you a ride to the ferry."

Peggy replied, "That would be very nice, Dave," and her acceptance led to our first opportunity to be alone together. As I drove, we talked. We laughed often and I liked her sense of humour. I asked her if she'd like to go for a ride with me the following weekend, and she accepted.

Soon, I drove her to Victoria to stay with her Aunt Nellie McMillan and her husband, a blacksmith in town. This aunt wasn't a blood relative, even though her last name was the same. She was an old family friend from Argyleshire, in Scotland, Peggy's hometown. From this visit on, Peggy and I went for Sunday drives most every weekend, creating our own world within my car.

During the spring of 1939, King George VI and his consort, Queen Elizabeth, were touring the country, but we didn't care. We enjoyed the simple pleasures we found together. I discovered she was kind, but not mushy. She was very Scottish, and like most Scots I've known, not much for kissing. I liked the fact she was precise and demanding of herself, just like her father, Hector. He demonstrated his precision in his gardening. Peggy demonstrated hers doing housework and cooking. She enjoyed knitting and liked flowers, especially roses. She was not only good-looking, she was hard-working like my own mother. And she'd a sweet nature, willing to help those in need.

On our drives together, my fondness for her deepened, but marriage was still the furthest thing from my mind. When I'd come to BC, I'd dreamt of staying only long enough to get well, then I planned to travel on to New Zealand. Only the Depression and the lack of money stopped me. Though I was fond of Peggy, I still had this dream of travel in the back of my mind.

But the politics of the world were changing, and I would have to change with them, like everyone else. The Depression was soon to be ended, by World War II.

On September 8, 1939, I'd just come in for lunch from a field I was ploughing. The radio was on loud and everybody was crowded around, listening. The other men's faces looked tense as I walked up to hear their news.

The Canadian parliament had just declared a state of war between Canada and Nazi Germany. By the end of September, fifty-eight thousand Canadians had joined the army, but none of them were from Colony Farm. Some joined for patriotic reasons, but many joined to get the army pay of $1.30 a day plus free room and board, clothing, and medical and dental care. After the years of depression, the army looked pretty good to those who had lived in relief camps, doing hard labour for twenty cents a day.[1] But the army had no such appeal to those already working steadily, like me and most of the other men at Colony Farm.

It bothered me when I realized that, for eight months after the declaration of war, no one from the farm enlisted in the army. I thought about it, then I decided to join up, but not for the pay. Like many Canadians, I joined partly for patriotic reasons, because I thought our participation would protect not only Canada but also Scotland. I also thought the army would give me a free trip home to visit my parents—an impossible expense during the Depression. As well, the traveller in me wanted a chance to see Europe. Those are the reasons I decided to enlist.

So in May I went to Geordie McMillan and said, "There's nobody at the farm joining up, so I think I'll volunteer. I'm told I'm guaranteed to have my job back when I return or when the war's over."

Geordie replied, "If that's true and you're going, I'm going with you." His decision was that simple. That was how the two of us became the first ones from Colony Farm to join the war effort. After our enlistment was announced, about seven other men decided to follow us, hoping to join our unit. (Geordie became a piper and went overseas to all the places I dreamed of going, playing the pipes during the war.)

The hard part was breaking the news to Peggy. When I told her what I was doing, she didn't say much, didn't try to talk me out of it, but I could see by her face that she didn't like it. Because I didn't think of myself as settled yet, there were still no promises between us. But suddenly I realized that I didn't like the idea of leaving Peggy alone to face a difficult future with her little girl. I thought, *I could get killed in this war. If Peggy and I get married before I leave, then if I die, at least she'll get my pension.* This was what made up my mind.

On impulse I said, "Peggy, let's get married. I know I'll be going overseas, but we'll get married before I go."

She smiled a wee smile and nodded to say "yes," and with those few words our future together was settled. My life as a bachelor soon became a memory.

My proposal may not have sounded very romantic, but I didn't think I had to be that way. In Scotland, for most of the people I knew, love wasn't the most important reason for entering a marriage. The people in my village got married because they liked each other all right, but I never saw a lovey-dovey Scottish couple. Most farm men married neighbour's daughters who they admired, but they kept their admiration pretty much to themselves. For most, admiration was inspired by a woman's talent as a good cook and housekeeper, and by her ability to work hard.

Once married, what happened between a man and his wife never went outside the house. As far as I was concerned, when people said "I love you," that meant sex, and sex was never mentioned in mixed company. In Scotland, if some young fellow was caught having sex with a girl, he got a goddamn good beating.

When my father talked to my mother in front of us children, it was "Jane," and she called him "Joe," as in "Joe, will you speak to these boys." Children were cuddled, but adults weren't.

Once children were in school, even their cuddling was forgot about. We had to toe the bloody line after that. I guess that's why I was so comfortable with Peggy. She understood what was meant by my proposal, even though it wasn't mushy. We planned a small June wedding in the vestry of the United Church, in Victoria, and I was permitted a week's leave from the army.

On the day of the wedding, so many friends arrived at the church that the minister said, "We might as well go into the main chapel because there's not enough room in the vestry."

As far as both Peggy and I were concerned, the wedding was just grand. I wore my uniform, of course. Peggy's brother Geordie was my best man and Peggy chose a friend from her work to be bridesmaid. Peggy wore a short pastel-coloured dress, not white of course. Peggy knew wearing white would cause a scandal, because she wasn't a virgin. Most of our close friends were aware she had a daughter.

Back in Scotland, both Peggy and I remembered instances where girls who weren't virgins defied convention and wore white. For punishment, these women had to stand up in church in front of the congregation and be lectured, deliberately shamed by the church elders . . . and some of those elders might have been the ones who led them astray, the bloody hypocrites! Things weren't as bad in Canada, but society was still very judgmental. So Peggy didn't wear white.

After the ceremony, we all went back to the home of Peggy's aunt and uncle in Victoria. A friend of ours, whose last name was Jones, had come to the wedding carrying six bottles of illegal scotch in his suitcase. He could hardly wait to get the ceremony over and get on with the celebration, but we told him to keep the booze hidden until after Peggy's parents had left. Until then, it was tea and sandwiches. Peggy's mother was against liquor and we respected that.

After she left, it was a great wedding celebration, made even more meaningful because all of us knew Peggy and I would soon be parted by the war. We both felt happy, surrounded by people who cared about us and wished us well. I discovered there were more people than I realized who liked me. I remember thinking, *maybe it's because I've been carefree up until now.*

Our honeymoon was very short. First we spent a day in Duncan, where we attended a big Sons of Scotland picnic in the park. Many other friends from New Westminster and Colony Farm were at the picnic, and when they met Peggy, I could tell they liked her. Then we travelled to Nanaimo, spending the night in a hotel. The next day we caught the ferry, returning through Vancouver to Colony Farm to tie up loose ends and say one final goodbye to our friends at the farm.

When we arrived, my good friend Erskine the blacksmith— whom I'd helped shoe horses when he had a bad back—shooed a couple of the other fellows out the door towards the fields, but I didn't think anything of it.

"Come on into the shop, you two," he said. " I'd like to talk for a bit, before you leave us." We chatted a while, with him asking us about our plans, where we were headed with the army, and general chit-chat. Before long, we noticed this large group of people coming into the blacksmith's shop, carrying something between them.

At first Peggy and I didn't realize they were coming because of us. We couldn't see what it was they were carrying, until they clustered all around us. Erskine had sent the two men to bring the other workers in from the fields.

The blacksmith was a big, jolly man but he wasn't used to making speeches, so he coughed nervously before he began.

"The long and the short of it is, we're sorry you're leaving us, Dave," he said. "But we wish you happiness in your marriage, and the good luck overseas to come back and enjoy it."

With that they presented us with a beautiful black stone mantel clock, with a gold-coloured face and an engraved inscription reading, "Presented to Dave Caldow from his fellow workers, on the occasion of his wedding, June, 1940." I was secretly touched, but I made a response that started everyone laughing again. Peggy and I treasured that clock ever since that day.

After the wedding, the army gave me basic training in Victoria for about four months. To have her near me, I rented Peggy a room in Victoria, but I saw her only on weekends. This was a strange way to begin a marriage, but we weren't alone in this strangeness. There were many other couples who married just after the husband entered the army, before he was shipped overseas. Prior to my shipping out, Peggy and I said good-bye to our old friend, my 1929 dark blue Chevrolet Coupe. I sold it for about five hundred dollars, adding a nice profit on the three hundred I'd saved from my Colony farm pay to buy it in the first place.

At the end of four months of weekends together, I boarded a troop train bound for Camp Borden in Ontario. I was bored with the long train ride, so one day, to create some excitement, I started teasing another Scot, one of the Red Cross men on the train, who had been a butcher in Vancouver. Just for fun, I blocked the door leading to the next car on the train and I wouldn't let him pass. He grabbed me and we wrestled each other. Somehow I slipped and fell, hitting my mouth on a seat, breaking my top plate of false teeth. (We didn't know what a toothbrush was, in Scotland, so I'd had no upper teeth since I was sixteen years old.) I felt very foolish being without those teeth for the rest of the trip!

As soon as I could after arriving in Borden, I got myself some army teeth. The army dentist convinced me I also needed a partial bottom plate, so I told him to go ahead, not knowing the trouble I'd let myself in for. The damn partial plate didn't fit right. It was so tight, it gave me the worst headache. Yet, try as I might, I couldn't get it out of my mouth by myself. Finally one of the fellows in my barracks relieved my agony by taking a screwdriver to pry those damn teeth out of my mouth. I went back to the dentist, but in spite of his efforts, those teeth never were comfortable. I promised myself, the first thing I'd do when I got home again was get rid of those army teeth.

For the next couple of months, the army sorted us according to our abilities, giving us further training according to our assigned jobs. As I had enlisted as a truck driver, the army at first assigned me to be truck driver attached to the army police force, helping to keep the troops in order. Then somehow one of the captains heard I'd had some blacksmith experience. He needed a blacksmith, so he asked me if I'd like to switch to fill the position. At first I refused. Then he pleaded with me, telling how much they needed someone, stressing I'd earn an extra twenty-five cents a day. His offer made me think, because the twenty-five cents extra would give me more spending money. Most of my pay at that time was sent directly to Peggy. In the end, I gave in and agreed to try passing the blacksmith test.

Back in the barracks, I told another Scotchman about my decision and about the upcoming test. Having taken it and failed once himself, he told me, "You'll probably have to make a link for a metal chain." Rummaging in his duffel bag, he produced some welding compound to put on the link, telling me, "This'll keep the tempera-ture low enough that the metal won't burn and melt too fast."

"Thanks for the advice," I said, "but to tell you the truth, I don't really care if I pass this blacksmith test."

Nevertheless, on the day of the test I put his compound in my pocket and walked in the sunshine across the parade square towards the smithy, a big tricky Swede—a sergeant who knew his business. He was giving me the test. In the heat of his shop, he walked behind the big black anvil, then said, "Caldow, come over here. Are you a blacksmith?"

"Well now," says I, "I won't know that for sure until ye finish with me, will I?" The sergeant laughed, and I thought *at least he's got a sense of humour.*

With that he turned his back, bent over and began effortlessly dragging a big logging chain, throwing it up on the anvil. He paused, looking first at the chain and then over to me.

With a challenge in his voice he said, "Okay, let's see you put a link in that." He handed me a little ballpeen hammer you'd use to break candy, and a cold chisel a half-inch thick.

I said, "What are these tools for? What are you trying to do? Insult me? Christ, I didn't think I looked that green! You know as well as I do, nobody could put in a link with those tools! Give me some decent tools and you'll get you're link."

Smiling a little, he handed me the right tools. "Well," he said, "Go ahead, but first remove this link, here." I looked at the link he pointed to and saw it was made of steel, not cast iron. I told him so. Then, without much difficulty, I removed the link.

I suppose directing me to a steel link was another trick of his to test whether I knew my business, but I thought, *he's not the only one with tricks up his sleeve.* As I worked, I quietly pulled out the welding compound my friend had given me and smeared it on the metal. The compound helped and when I finished, I felt satisfied I'd made a perfect link for the chain. I handed it to the sergeant, who nodded but said little in response.

"Okay Caldow," he said, "now show me if you know how to tem-per a chisel." He was shoving the chisel towards me as he spoke.

I replied, "It's a long time since I've done one of these, but I remember how to do it." And I did.

The sergeant just grunted when I finished and handed him the tempered chisel. As a result, I left him not knowing whether I'd passed the test. I smiled later, when I learned that Swedish sergeant gave me a first-class blacksmith's ticket.

When I told my friends who knew me from Colony Farm, they just laughed. "You, a blacksmith?" they said, as they shook their heads. Of course, they knew nothing about me helping Alex Hutchinson, the blacksmith in Claresholm, years before.

On the steps of the post office at Truro,
Nova Scotia

Peggy came out for Christmas and I was given a short leave. We spent it revisiting Macdonald College, in Ste-Anne-de-

Belleview, Quebec. We were guests of the new superintendent's family, Jim Houston, the son of the man who had been superintendent when I worked there. We learned his father, Jack, had been killed in a train accident, and I missed Jack's presence. Through conversation during the delicious Christmas dinner, I learned that the veterinarian, Dr. Conklin had also joined the army, but not as a veterinarian. In the years since I'd seen him, he'd returned to school and become a medical doctor, and a very good one too. I was enjoying visiting with old friends again, but that Eastern climate soon took the joy out of it for me.

I was sitting beside their warm fire and realized I was shivering. Peggy borrowed a thermometer and took my temperature— and it was only 92. My friends called to the doctors, and they ordered me into hospital. I spent most of my leave in a hospital bed. Bronchitis had returned, the same as it had been the last time I'd been in the East. I thought, *Christ, this climate would kill anybody!*

When I returned to Borden, I didn't tell the army I was sick. I didn't want them to know anything about it, because I wanted to go overseas. If they found out, I knew they'd either keep me in Eastern Canada or ship me home. I was soon ordered to Deburt, Nova Scotia, but not before I received some sad news from relatives, in a letter that told me my mother had died of old age, on Valentine's Day, 1940. She was seventy-two so the news was no shock, but still I was saddened and wished I could have seen her again. I'd never been able to afford either the time or the money to return home to visit with her, what with the Depression and all. She didn't write me often, but I had always enjoyed her letters whenever the postman brought them. Now I knew there would be no more of them.

In Debert, my sickness got worse, giving me everything from severe sinusitis to bronchial pneumonia. There was no way around it, I had to go into Camphill Hospital for a time. The first time they released me from Camphill, I passed out on the parade ground, unable to breathe. The men picked me up and carried me into the hospital again.

In the examination room, a doctor everyone called "the old Jew" was there, examining another man ahead of me. He was lecturing the other man, saying,"There's nothing wrong with you! Don't waste my time pretending you're sick, just because you want to get out of the army!" With that, he turned to me and said angrily, "What's the matter with you?"

I said, "Nothing. Just give me a couple of aspirins."

Peggy, when in Truro with me.

The doctor put his stethoscope on my chest, listened to me breathe, then turned back to the other patient and said—still angry—"You're telling me you're sick and there's nothing wrong with you. Yet here's a man half dead who just wants aspirins!"

Panicked, I thought to myself, *Christ! Half dead! He says I'm half dead!* With that the doctor ordered me right into the hospital, again. This time I didn't argue. I was in Camphill hospital for two of the three months I spent in Debert. Though I'd much rather have been out learning to soldier, it was the best hospital I was ever in. They served us real home cooking, prepared by the women of the town.

This time when I was released, they gave me a letter saying, "This man is exempt from military service due to ill health." I was ordered to return to British Columbia on the following Wednesday, on a special train for those who had been lowered from "Category A—Fighting Men" to "Category Home Defence." This meant my train would have no beds, even though I was still sick. As well, this quick departure left me no time to arrange for Peggy to return with me. She would have to return on her own. The orders upset me, but I didn't think there was much I could do about them.

I ran into the old doctor again, just as I was on my way to tell Peggy.

After saluting him I said, "Doctor, you buggered up my return home."

"What's happened?" he asked.

I told him, "I've to go on the Category train."

He shook his head and said, "Follow me!" With that, he marched me into the office and demanded I be given papers to assure me a bunk all the way home. He also arranged for a time extension until the following Friday, got me a meal ticket, as well as passage for myself and Peggy on a passenger train. I was very relieved.

I thought to myself, *This old doctor isn't too well-liked, not only because he's gruff, but because he's a Jew.* Yet, he's intelligent, straightforward and honest with his opinions. I thanked him and knew I liked him, and not just because he helped me.

At this time, I was asked if I was going to apply for an army pension. I'd heard rumours about tricks that other men—also exempted from military service—were using in order to get pensioned. In my barracks, I knew a guy from Saskatchewan who bragged he'd made up a story about how he'd been kicked in the head by a German prisoner, just so he could get a pension. His head injury was the result of a fall in camp. I thought this behaviour was dishonourable.

Personally, I was disappointed, and not only because I lost my chance for a free trip to Scotland. I felt I'd missed the chance to protect my country. So I answered the question with, "No, I'm not going to apply for an army pension, because I don't feel I've earned it." (I still feel I did the right thing not applying, even today.)

When I climbed aboard the train in Nova Scotia, I was simply relieved that, true to their word, the British Columbia Government and the people at Colony Farm would welcome me back, giving me the same job I'd had when I left—truck driver. But that job didn't last long.

CHAPTER TEN

Fatherhood and a Different World

The youngling Cottagers retire to rest:
The Parent-pair their secret homage pay . . .
—Robert Burns, "The Cotter's Saturday Night"

After a year in the army, I found the world of Colony Farm very different from the one I had left. Some of the changes related to my work and some were deeply personal, but most of them eventually proved to be for the better. In one short year, the depression had become just a haunting memory. Suddenly, more civilian jobs existed than there were men to fill them, due to the numbers joining the armed forces. I had also changed. My experiences had transformed me into a man of responsibility.

At work, I said what I thought. Oh, I wasn't full of myself, but I was confident, and that gave credibility to my opinions. As time went on, I saw some stupid mistakes made for a variety of reasons. Sometimes people were crooked and out for their own ends. I could see the problems in a big government-run institution, where many people must work together, even if they don't share similar points of view or knowledge. Those at the top didn't want to listen to practical advice from us workers. There's a lot of politics in a big place, and bad decisions made for political reasons sometimes drove me crazy. I don't know, to this day, why they didn't fire me more than once. I was free to speak out because I wasn't afraid, because I always knew I could find another job. My father and my years of farm experience had given me valued skills I knew were needed most anywhere. By the time I returned from the army, I was confident enough to survive the bureaucracy of Colony Farm.

At first, Peggy and I hadn't a house to live in. She temporarily stayed with friends while I continued to live in the show barn at Colony Farm. As soon as I could arrange it, I found Peggy an apartment in the Burr Block on Columbia Street in New Westminster. I came home mostly on weekends due to the length of the trip into town, travelling either on foot or hitching a ride with someone else. By this time Peggy was pregnant, so I was anxious to get us settled before the baby arrived. Not long after we moved, I was walking back to Colony Farm along the railroad tracks, feeling haunted by the thought that Peggy seemed to be keeping something important to herself.

She hadn't looked well when I left her, but when I'd asked her about it, she'd replied, "I'm fine, Dave. You be on your way now." But the further away from her I walked, the more worried I became. By the time I arrived at Colony Farm, I was sure something was very wrong with Peggy.

Immediately, I ran to the garage and borrowed a truck to use for the return trip to our apartment. On arrival, I ran up the stairs as fast as I could. Without knocking, I flung our door open and called, "Peggy?" But there was no answer. She was nowhere in any of the rooms. I ran back down the stairs, jumped in the truck, and drove like a wild man to the Royal Columbian Hospital.

At the hospital reception desk, I said, "I'm looking for my wife, Peggy Caldow?" I stood impatiently waiting while the name was checked.

I felt tremendous relief when the response was, "She arrived about an hour ago, Mr. Caldow. Please go up to the maternity ward. I'll let them know you're coming."

It seemed to take forever for the elevator to reach the correct floor. As I stepped into the corridor, the nurse greeted me with, "Mr. Caldow? Congratulations. You're the father of a fine, healthy baby boy."

"How's my wife?" was the first question I asked.

"She's fine," Mr. Caldow," the nurse replied. "Please come with me and you can see for yourself."

Relief came in a wave when I saw Peggy's smile as we entered her room. She was holding Geordie wrapped in a blanket. She still looked tired, but also relaxed and pleased.

"You have a son," she said softly. "Are ye pleased?"

"Of course," I said. "But Peggy, why didn't you tell me?"

Proud parents.

Without answering my question she smiled saying, "Don't you think he's in your image?"

All I could see was his face above the blanket. I looked closely at him. I was a bit disappointed to see he looked just like any other baby to me, but I didn't say that to Peggy. She looked so proud.

"Hold him," she said, so I did.

At that moment all anxiety disappeared and happiness flooded through me, but there was no way I could put my feelings into words.

"I'm so pleased you're both all right," was all I could choke out. Yet, my words were enough to satisfy Peggy. She understood. I sat with Peggy for the rest of the night and in the morning I went to work and proudly announced to everybody, "Peggy and I have a son."

During her hospital stay, Peggy had many visitors from the Sons of Scotland. We were lucky because a friend—the wife of Jim Watson the fire marshal—offered to have Peggy and Geordie stay at her house for the first three weeks of her recovery from Geordie's birth.

During this time, I located a small house for us to rent, owned by Erskine, the same blacksmith who had presented us with the marble clock when we were married. Though the house was small—located on River Road in Port Coquitlam, not far from the red bridge—Peggy smiled when she saw it. From the first, she kept that little house so clean, so comfortable, and always scented with roses. Peggy loved roses. Whenever she could, she filled our houses with them. I enjoyed having a home of my own after living so long in other people's places.

Not very long after Geordie's birth, our small family was increased by another member, but not because we had another baby. Bobby Caldow, my brother Bob's teenage son, came to live with us. Bobby was in need of a high school, with none to be had near his home. We offered to have him stay in town with us. His arrival made more work for Peggy, but even with Geordie so small, she agreed to welcome him.

Peggy's wee daughter Ethel remained with Peggy's mother, not because Peggy wanted it that way, but because I did. It seemed the sensible thing to me because everyone—other than close family—believed Ethel was Annie McMillan's daughter. (Even our son, Geordie, didn't find out the truth until many years later.)

Once in a while I couldn't help thinking, within the span of a year and a bit, I'd gone from bachelorhood to becoming the father of a baby boy and stand-in father for a teenager. I knew I'd reached a major turning point in my life and there was no going back. My single days were definitely gone forever.

Peggy's daughter Ethel, and Peggy's father with his namesake, Geordie.

However, being married had its compensations—one of which was Peggy's cooking. She made me scones after I bought her a new griddle. They were delicious, but of course, not quite as good as my mother's. But then, it's always difficult for a reality to compete with a memory. Peggy's shortbread was the best, bar none, though she didn't call them shortbread. They were "icebox cookies" to her. Many times I sat and watched her mix a pound of butter, with rice flour, salt, and sugar, then roll the mixture and

put it in the fridge overnight. In the morning, she'd cut the cookies and mark them with a fork, then I'd return from work to the
delicious smell of them baking. Our home and her food were
quite a welcome change from my bachelor existence and the food
in the farm kitchens.

Though our rent was agreed to be fifteen dollars a month, I
never paid Erskine a penny of it. I didn't need to pay him because
I worked for him on weekends and some evenings, helping him
to break and train horses, earning fifty cents an hour. By the end
of the month, instead of me owing rent, Erskine often owed me
money for the work I'd done.

One weekend Erskine brought me one of his smaller work-
horses, a black half-Clydesdale who had almost killed his driver
at the end of a furrow. The horse had been pulling a disc for plowing when he jumped and kicked hard enough to tip the heavy disc
over, landing it on top of the frightened driver. The fellow was
lucky, he survived with only a few bruises, but after this incident
the horse couldn't be worked. He began jumping and kicking
whenever any reigns moved across his back. The horse's behaviour was puzzling, until I examined the original harness and
found the big bent buckle, sitting centred on the horse's back. The
point of the pin on the buckle dug into the horse, causing him
pain. The pain had prodded his jumping. Trouble was, the jumping had become a habit, occurring even with a new harness,
whenever a driver pulled on the reigns to make a turn. Now that
I understood the reason for the horse's strange behaviour, I
planned the cure to break the habit.

What I did sounds rough, but it was better than having the
horse destroyed. First I led the horse, in a comfortable harness,
into a newly ploughed field. Then I attached two straps, each with
a metal ring, around the horse's front pasterns (ankles). I placed
a third strap circling his back and belly, having one metal ring
attached below the belly and one ring on each side. I threaded a
rope through the three rings on his front pasterns and belly, making the rope form a w. Then I threaded each end of the rope
through the two rings on the sides of the belly strap. This way,

when the horse jumped, I could pull the side ropes from behind the horse, drawing both pastern rings back towards the belly ring, collapsing the horse's front legs when he was in the air. Of course, I made sure we were doing this in a soft field so he wouldn't be seriously injured.

At the first turn the horse jumped. I immediately pulled the rope and he went down—whomp—into the dirt. The surprised horse snorted and shook his head, trying to shake the dirt from its face. He didn't like the indignity one bit! It only took a couple of falls. The sorry horse soon stopped his jumping and kicking in harness.

To be certain the animal was cured, Erskine and I worked him every night for about two weeks, making sure nobody would again be hurt. Erskine then sold the horse to McGavin's Bakery. That old horse led a good life for quite a few years, pulling a baker's delivery van.

Renting the house from Erskine proved to be one of the best deals I ever made. As well as the house, he allowed me to use some of his nearby land. On it, I raised a heifer bought from Spenser Stroyan, the herdsman at Colony Farm. Spenser wanted rid of the heifer because she had injured her legs on some cement at Colony Farm, and the injuries were taking too long to heal. Once she was alone in our pasture, she healed just fine. Later, I added six calves, which I raised to ninety pounds and sold for veal. I also bought two young pigs, raised them to two hundred pounds, then sold one and kept the other for our own food. I'd get up at 5:00 A.M. to milk the heifer and feed the animals, then Peggy would make me a big breakfast before I'd leave for work at Colony Farm. Working all day, and often in the evening and on weekends, I'd little time for leisure or for spending time with Geordie or Bobby. However, the deal with Erskine meant that Peggy, Geordie, Bobby, and I made out just fine, even though we didn't have much money, starting out.

I was more and more pleased with my situation at home, but I was less and less pleased with my situation at work. Things weren't right in a number of ways. Though I was still respected as a hard worker, when I tried to speak out about needed changes, it seemed to me nobody listened. The lack of response didn't stop my speaking out, though. Even with the police, I spoke my mind.

Like the time I was returning a load of pig carcasses to the kitchens of Colony Farm. I'd covered the meat with two big freshly laundered white sheets protecting it from dust and germs, as had been the custom since I'd started working there. A police siren pulled me to the roadside. A big gruff policeman ambled up and told me I was no longer allowed to transport meat in an open truck. He told me to tell the people at Colony farm to send a covered truck. I faithfully carried the message back, but no changes were made. The policeman stopped me two more times repeating the same orders each time.

By the third time, he was pretty irritated, but I was in no mood to take anything from him. I told him, "Look, I'm not hauling any more messages back to Colony Farm. It's not doing any good, because nothing changes. If you want them to get your message, you bloody well better tell them yourself."

The policeman didn't like what I said, but he must have acted on it. That afternoon I was told that the management at Essondale would now send their closed truck to haul the pigs. We at Colony Farm would no longer haul them. My speaking up had created one change, but it was annoying that it was necessary. This bit of success wasn't sufficient to make me happy with my working life. By the time a year and a half had gone by, I was discouraged enough that I was almost ready to quit.

My most troubling problem was, I didn't like working with the assistant superintendent, Art Hay. Oh, he didn't do me any direct harm, because I wouldn't let him. But I felt sure I couldn't trust him.

More than once when I'd gone to him with a request for supplies, he'd turned me down with the words, "We can't do that for you now, Dave, but when we get rid of the 'old man' (meaning Pete Moore) things will be different." It was the way he'd say it and wink that made me angry. I felt disgusted. Pete Moore had hired Art Hay, just as Pete had hired me. Art was bad-mouthing the man who had done him good.

No sir, I thought, *I want none 'a this.* Art Hay, yer a sneaky bugger, out for yourself. I'm sure you don't care who you'll be hurting in your drive to gain power.

I was even more upset when I overheard him encouraging the business manager to be his cohort, to support him in his undermining of Pete Moore's position. This was the last straw.

I went to Pete and told him, "I can't abide working with Art," nothing more.

Pete replied, "Hang on for just a little, Dave. I think I have something coming up that might interest you. Give me a few weeks to work it out."

True to his word, Pete offered me the chance to go to the Lizzmoil Farm at Hatzic, near Mission, by way of guaranteeing the performance of some cows purchased earlier from Colony Farm. Colony Farm took great pride in breeding champion high-production cows, yet the ones at Lizzmoil weren't performing. I listened to Pete describe the problems the dairy-cow purchaser was experiencing.

"Dave, if you accept, you'll be working for a millionaire named Andrew Houstoun, for as long as it takes you to improve his cows' milk production."

I said, "That sounds to me like a pretty good improvement on working with Art Hay."

At home, I talked to Peggy and she agreed with me, so I accepted the challenge at Lizzmoil Farm.

Just before we left the Colony Farm herdsman (Spencer Stroyan) gave me one more task. In so doing, he introduced me to the agent of my own catastrophe, to come in the following year. Spencer had me drive with him to a farm up near Deroche, (located past Mission and Hatzic, on Nicoman Island). We were to pick up a Holstein dairy bull the day after it had been involved in a tragedy.

A farmer had leased the Holstein from Colony Farm, but must have felt as if he owned it, because he'd had it ever since it was a year old. He must have fed it well, because the bull was a big bugger. His horns had been removed, but his head alone probably weighed close to 170 pounds—enough to crush anything he pounded with it. Maybe the farmer just wasn't wary enough because he thought he knew the bull too well. At any rate, he learned the hard way this bull wasn't to be trusted. When we arrived at the farm, the farmer's sons told us what happened.

Early in the morning, as was their custom, the two boys and their father had gone into their barn to do the milking. The father then left the boys, going outside to bring in more of the loose cows. His sons said their dad must have seen the bull with cows he didn't want bred, and probably went into the field to separate them. When their father didn't return to the barn, the boys assumed he'd gone up to the house ahead of them. But they were wrong.

As the boys walked back for breakfast, they noticed the cows in the field behaving strangely, all milling around. Filled with dread, they pushed their way through the nervous animals. In the middle of them, the horrified boys found their father lying dead, his chest crushed, with the bull standing triumphantly over him, sniffing him. After they got over their shock, the boys figured the bull must have swung his huge head, battering their father. The farmer didn't have a chance.

The first thing Spenser and I did was rope the beast and tie him to the wall of the barn. Strangely enough, the bull gave us no trouble. Next we put a ring in his nose, and this also was easy. We were helped because the animal already had a hole in his nose from a small calf ring, left unreplaced by the farmer.

In spite of the horrible death this bull had caused just the day before, he seemed placid enough. But he was just waiting. I knew it. I could feel his meanness and see it in his sneaky eyes. It took a while, but the next time I tangled with him, this beast damn near killed me, too.

CHAPTER ELEVEN

Hatzic

Ye flowery banks o' bonie Doon,
How can ye blume sae fair . . .
—Robert Burns, "Ye Flowery Banks"

When Pete introduced me to Mr. Houstoun, I was impressed with him, and not just because he was a fellow Scotsman. By reputation, I knew him to be a real gentleman who treated people with respect. Houstoun told me he'd welcome Peggy, Geordie, and Bobby coming to his farm. He explained he lived at the Vancouver Club during the week and came out to Lizzmoil Farm only on weekends. Mr. Houstoun was a prominent man—a millionaire who owned two or three fish canneries—as well as the McLennan, McFeely, and Prior chain of hardware and household goods stores. In the late 1920s, Houstoun had taken over as president of the chain of stores because he'd loaned money to McLennan and McFeely, but they hadn't made good on their debt.

He bought Prior's Hardware in Victoria because he wanted to employ that store's manager, who was a hard-headed business man. Houstoun thought employing him would help his business. But I never liked the man, who seemed full of himself. I wouldn't have worked for him for a minute.

When Peggy and I were making arrangements to move, Houstoun told us to sell all our furniture from the house in Coquitlam. He said he'd give us a letter entitling us to buy new furniture in his stores when our job with him was done.

He said, "The letter'll allow ye go to Mc and Mc's (the common name for Houstoun's stores) and buy *new* . . . at my

141

expense." Since Pete Moore had told me Houstoun was a man of his word, I didn't get his promise in writing. I took it on trust.

I asked Houstoun how he'd become a farmer. "I needed the farm to give me something new to take up my attention," he said. His answer surprised me. I was impressed that at seventy years old he'd been looking for more to occupy himself.

"By the way, Caldow," he said, "Your wife Peggy can work as housekeeper for Lizzmoil, if she's willing."

I said, "Listen, Mr. Houstoun. I'm not willing to bring my wife up to Hatzic to have you work her to death, cooking for all the fancy friends you'll be entertaining."

Houstoun laughed and said, "Listen Dave, I've come up there to get away from all those people. Your wife won't have any trouble with me overworking her, I assure you."

Later, Houstoun told me he was glad Peggy had decided to work for him. He liked the way she filled his house with roses, cut from his prize beds. But he never ever pushed her to do more than was fair. He kept his word.

As soon as we could, we packed the few personal things we took with us. I told my nephew Bobby—now 16 years old—he'd be helping me with the farm chores. Later, these chores turned a profit for Bobby because, when Houstoun saw him helping me, he insisted Bobby be paid for his work. (It wasn't long after this that Bobby graduated high school, then went on to university.) Geordie was still too small to be helping, but both boys were excited about the move and wanted it.

The turnoff to Lizzmoil Farm was on the way to Aggasiz, past Mission and just past Hatzic at the top of a hill, near a little store and post office. On a clear day, we turned into the driveway of our new home. We were greeted by Mr. Houstoun's pride—the perfume and colours of a large bed of red, yellow, and white roses blooming in front of the house. Sunshine made the whole 160 acres—including tilled fields, pastures, and lake—beautiful and welcoming.

Over the next few days, I met the three hired men, as well as the gardener and the cowman. All of them were paid by the hour to do the farm work, with no long-term guarantees.

As soon as I could, I looked over the cows because their poor performance was the reason I was here. I thought, *In this beautiful setting, what could possibly be keeping you cows from performing like the champion stock you are?*

Part of the answer to my question came suddenly, about a week later, as I came through the door into the milking barn. I couldn't believe my eyes, when I surprised Houstoun's cowman furiously beating one of the cows with a hose from one of the milking machines.

I saw red! Without asking for an explanation, I ran to him, grabbed him by the throat and yelled, "If I ever see you hittin' another cow here, you'll be going down the road. What's the matter with ye man?" With that, I let him go.

The cowman was shaking when he answered, "She . . . she . . . she was kicking me when I was hooking her up."

"There's ways that work without hitting," I said forcefully. "If ye knew she'd kick, ye could've tied her feet or withers. There's no need for anybody hitting a poor dumb cow!"

I'll tell you, he heard me. He never hit another cow on Lizzmoil. In fact, before I left he'd turned out to be a pretty good cowman.

Two or three weeks later, when the first milk revenue cheque arrived, I realized just how badly these cows had been performing. Houstoun was only earning $150 for the milk from twenty-five cows, far less than he should have been getting. Immediately I changed their feed by adding minerals, and purchased mixed feed for their diet. Before I came, they'd been fed only the grain grown on the farm. Then I set about to improve the growing of Lizzmoil fodder by putting three hundred tons of lime on the field that produced it. At the end of seven months, milk revenue was up to some $980 for the milk of the sixteen producing cows. A big improvement. My father's Scottish training once again helped me make a big difference to the animals.

❧

Not long after our arrival, both Peggy and I had the same worrisome idea that something wasn't quite right with Mr. Houstoun. We were surprised when we noticed that the first thing he'd do when he arrived home was pour himself five or six ounces of straight whiskey in a big glass, then fill it with cream and drink it.

He told Peggy, "Every weekend I want you to be sure to have a pint of cream and fresh orange juice ready for me." He'd mix both with alcohol and down the lot. Yet we never saw him the worse for it. In spite of his drinking, he was a good old scout and we grew to like him very much.

Being a Scot himself, he especially enjoyed Peggy's scones, served with her homemade jam. When he'd go into the Vancouver Club, he'd ask her to bake extra, then take some to share with his friend, a man named Johnson. Houstoun told me one day, he and Johnson were the only two close friends left out of a group of seven who had joined the Vancouver Club together. "That's a problem that comes with age," he said, "losing your old friends."

But Houstoun had some non-human friends to comfort him. I enjoyed working and training his two favourite dogs, a Labrador retriever and an English setter. Houstoun said, "I brought the lab to Canada from England, because they were killing dogs off, there. During the war, there wasn't enough food for people, so the dogs had to go." Houstoun had long been raising dogs, so he and I shared more than one common interest.

About seven months after we arrived at Lizzmoil, Mr. Houstoun—by this time, Peggy and I privately called him "the old man"—got very sick. When he had recovered a little, we found him cleaning out his room and burning all his clothes.

I said, "Mr. Houstoun, are you all right?"

He nodded and I said, "You know, you drink too much. Are you drunk now? Is that why you're behaving so strangely?"

"Dave," he replied, "ten years ago the Mayo Clinic gave me a year to live. I've got cancer. When I returned from the clinic to Vancouver, I thought I might as well enjoy myself during my last year. That's why I started drinking. Oddly enough, it didn't seem to harm me. I've lived ten years instead of the one that was predicted, but I'm near death now."

"I'm so sorry," I said, but he would have none of my sympathy. He told me to phone a friend in Vancouver.

"Tell him to come and get me," he said. The friend came right away. He asked me no questions, knowing Mr. Houstoun wouldn't have asked unnecessarily.

Peggy and I stood watching them drive away. We felt helpless, sad, and a little lost. We waited anxiously for word about his condition, but not for long. The phone rang and we learned his Vancouver doctor sent him to a cancer specialist in Seattle, where an operation was quickly arranged for the next morning. Mr. Houstoun died within hours after the operation.

Strangely enough, Pete Moore and Art Hay came up to the Lizzmoil Farm the day Mr. Houstoun left for the hospital. They were wanting me to return to Colony Farm because the herdsman had left, and they came to offer me the job. They told me—if I accepted the offer—they'd see to it I had a house built for my family. I was pleased, and I knew Peggy would like the idea of returning, too. I said I'd come, but only after I'd arranged things properly for Mr. Houstoun's farm.

Some time later, a lawyer named Montgomery—he said he was a brother of the famous General Montgomery—came to Lizzmoil. He was finalizing plans for Mr. Houstoun's estate. As Peggy and I chatted with him, he told us about the carpenter and his helper, who had been working on the farm. It seems they both had gone to Mr. Houstoun's funeral.

Montgomery said, "Would you believe it? Both men had the nerve to try and charge Mr. Houstoun's estate for their trip." Peggy and I were disgusted. When next I saw him, I told the carpenter exactly what I thought of him and his charges. (You can be sure, I paid him no compliment!) Montgomery also told us Mr. Houstoun had written to his brother, telling him I'd probably stay on at Lizzmoil, after he died.

I said, "I'm sorry, Mr. Montgomery. Frankly, I don't see any future for us here, now that Houstoun's gone. I think Mr. Houstoun's family will probably sell the property, so I've made up my mind. We'll be returning to Colony Farm."

Andrew Houstoun Claimed by Death

Andrew Houstoun, prominent in Vancouver business circles for many years, and president of McLennan, McFeely & Prior Ltd., died Saturday at Vancouver General Hospital.

He was first identified in business circles in the canning industry and later went into partnership about 1900 with E. J. McLennan and R. P. McLennan in the hardware business. Mr. Houstoun had been president of the company since 1928.

A member of the Vancouver Club, where he made his residence, he was an active sportsman and was well known as a hunter. Of late years he spent much time at his farm at Hatzic in the Fraser Valley where he was experimenting with dairy farming.

He is survived by two brothers in England.

Newspaper clipping of Andrew Houstoun's death.

"I'm sorry to hear that, Dave," he said. "I know the family will be disappointed because they were counting on retaining both of you. But I wish you all the best." Then he reached into his black leather briefcase and handed me Mr. Houstoun's letter of permission to purchase new furniture for our soon-to-be-built new home at Colony Farm. Peggy and I were feeling pretty good when we saw that—yes, we were.

We had the grounds for good feelings, believe me. The country was not long out of the Great Depression, with people still joining the army to get work. I'd been given a promotion to herdsman at Colony Farm. Added to that, we'd just been promised a new home and new furniture. Though we were both still sad at Mr. Houstoun's passing, things were beginning to look pretty good for our own future.

We were both glad I'd followed Pete Moore's instruction to wait for him to find a place where I could be free of Art Hay's conniving. Because of luck and friendship, we'd had the chance to know Mr. Houstoun, a gentleman of high principle, a man of his word. My father would have agreed, those are qualities worth knowing in any man, rich or poor.

CHAPTER TWELVE

Return to Colony Farm—Herdsman

Now gawkies, tawpies, gowks and fools,
Frae colleges and boarding schools,
May sprout like simmer puddock-stools
In glen or shaw . . .
                    —Robert Burns, "To William Creech"

Our new house would seem small by today's standards, but in 1944 it was an ample home, attractive with its white board siding. We were pretty much happy in it. Bobby continued his schooling and Geordie did the usual things that wee boys do. Peggy enjoyed decorating the house inside and out and was fussy about the keepin' of it. She made me take my shoes off in the basement, which made sense to me. She'd let neither me nor nobody else muddy her floors, not without them hearing about it. She worked hard to make the outside of the house as pretty as the inside, planting green shrubs and roses growing in front and climbing over the fence. Our home was my refuge, orderly and predictable—different from either my workplace or the events going on in the world.

Everywhere, the Second World War was on everyone's mind. In the evenings we listened anxiously to the radio news from the front. Though I was certain the Allies would win, I wondered how long it would take them, and how many lives would have to be lost before the German defeat.

Even if there'd been no deaths, the war was already creating a shortage of men available to do needed work, with so many in the armed forces. This shortage meant I had to deal with more than the usual herdsman's challenges. As well as supervising and seeing to training, I frequently had to do some of the more routine

Note to chapter 12 is on page 241.

jobs myself, such as collecting semen from the bulls for our insemination program. That's how, one February morning, I found myself again facing that murdering Holstein bull that me and Spenser had picked up from Deroche.

I was still wary of the beast—but not worried enough to prevent me from working with the bugger. He'd been a calm animal for many months, since his return. I later found out that his record for calmness was deceptive because he'd been left to himself as much as possible. The other men had shied away from him, and no wonder.

One morning I walked up to his stall, snapped the rope into his nose ring when he lifted his head, then had a couple of farm labourers help me walk him out to the exercise pen. When I checked him over, I realized he'd grown bigger since we'd picked him up. As I moved around him, his calculating eyes watched every move I made.

"Ye're such a big old fella," I said to him, chuckling. "No wonder you're on the breeding list. Come on with me and I'll introduce you to yer lady."

I laughed as I said "lady," because I was about to encourage him to mount a "dummy" cow, a contrivance we used when collecting semen. It consisted of an iron frame about the same size as a cow, constructed strong by Erskine, the blacksmith. On top of the iron frame we placed a mattress for softness, then covered it with cowhide, to make it more appealing to the bulls. But in spite of the fact I'd walked this beast around to the back end and muttered some encouraging words to him, the dummy cow wasn't appealing, I can tell ye. This old bull just snorted at it, then glared at me, with his breath forming steamy streams as it left his nostrils in the cold morning air.

I stepped back a bit to give him a minute to think about it and, because I'd been smoking off and on since I'd left Claresholm, I suddenly felt the need of a cigarette. I turned my head away from him for a moment, to light my cigarette out of the breeze, and that was my big mistake. That's when the bull hit—not the dummy cow—but me.

The next thing I knew, I was ass up—with my head in a corner of the pen and one of my legs bent up my back and over my shoulder. That bull had me eating dirt and I was helpless. I was frozen, by both fear and by the force of his blow to my stomach. I thought of the dead farmer . . . and then I thought, *Christ, the bastard's got me too!*

Farm Worker Mauled by Bull

NEW WESTMINSTER, Feb. 10. —David F. Caldow, employed at the Colony Farm, Essondale, escaped serious injury Thursday when he was mauled by a large Holstein bull he and two assistants were moving from one pen to another.

Caldow is in Royal Columbian Hospital with an injured hip, back injuries, and head bruises received when the bull suddenly butted him into the corner of a pen. He was rescued by the assistants who dragged the bull away.

The paper got it wrong. It was actually an engineer working nearby who stuck the bull with a pitchfork and rescued me.

But the bugger wasn't finished with me yet. I felt the ground shudder from his weight as he snorted and come at me again. I could see the toe of my one boot close enough to my mouth to bite it. Then I felt waves of pain flooding up from my right leg— which I later learned was pulled out of joint and up my back, with the muscles torn down through the groin. Pain screamed in my back from two fractured vertebrae!

I'm dead! I thought . . . then nothing. I was out, thank God!

That devil of a bull either stomped or butted my head hard enough to fracture my right eyebrow, forcing a sliver of bone down against my eyeball. I could feel the pain it caused as I woke up much later in hospital, lying on my stomach after the doctor had worked on me.

Good old Dr. Robertson was telling me, "Take it easy, Dave. You'll be all right, just lay still and rest." He gave orders not to move me for two days, along with instructions to give me a glass of brandy every day.

That brandy is the part I like to remember best. It's a better memory than the sight of my stomach looking more and more like a big piece of purple liver as the days passed, or the memory of the pain of the procedures that followed.

Dr. Robertson told a friend of mine about the difficulty he had getting my hip back in place. When I was under anaesthetic, he had a team of seven nurses and other medical people pulling on it. Even with that many people, it wouldn't go back. Six times they pulled, but it didn't move. Everyone but Dr. Robertson wanted to give up.

To encourage them to keep trying Dr. Robertson said, "Now look, this man's a Scot like Robert the Bruce. Robert the Bruce didn't defeat the British until his seventh try. So let's give it one more go."

Thank God they did and were successful. Dr. Robertson told my friend I had the strongest leg muscles he'd ever seen—and he was the doctor for the lacrosse team, so he'd known a few strong men.

Hell, when I think of it, why wouldn't I have strong legs? Ever since I was a small boy, I'd been walking behind plows. Before I'd left Scotland, I'd been lifting sacks of grain that weighed well over a hundred pounds . . . for years I'd laboured like a bloody work-horse. In Canada, nobody asked me to lift more than hundred-pound sacks, but that was more than enough to keep my legs strong, with all the other physical work I did.

I grew to really like Dr. Robertson, and so did Peggy—especially when we were invited to dinner with Mrs. Robertson doing the cooking. Even Geordie would go to Dr. Robertson's office without complaining—and that was something—because when Geordie was small, he was afraid of doctors.

Geordie's fear came from an experience he'd had while we were living on Houstoun's farm. He was not much older than four when he decided to help Peggy wash clothes. In those days there were no automatic washers. Clothes were agitated in the machine's tub, looking very similar to the machine tubs of today. But instead of having the water spun out of the clothes, the women had to lift them by hand, feeding them through two electrically rotating, hard rubber rollers. Each was about the size of a rolling pin, and forced hard together by clamps inside the machine. Like most children, Geordie would stand fascinated, watching the dripping clothes being pulled into the rollers, then fed out the other side, flattened and wrung to dampness. The clothes were then dropped into a clothes basket and Peggy would carry them outside to be hung to dry in the sunshine, attached by wooden clothespins to the outside clothesline.

On the day Geordie got hurt, Peggy had stooped to pick up something as she was doing the wash. Geordie grabbed a handful of wet clothes from the tub, then pushed them with his wee hands, up towards the rollers. Trouble was, he didn't let go when the machine took hold. The clothes tightened around his fingers and pulled his one hand through the rollers with the clothes. He screamed bloody murder, causing Peggy to jump up and hit the release lever, but it was too late. Geordie's thumb had caught upright against the roller, preventing his arm from being pulled through, but the skin of his thumb and palm was bleeding. It had been badly torn by the pressure and rubbing of the rollers and clothes.

Running, Peggy carried Geordie out to me with him still screaming. We bundled him into the farm truck and took him to a doctor in Mission—and what a callous butcher that doctor was. He stitched wee Geordie's hand without using anaesthetic. Geordie screamed all through it.

After that experience, our boy didn't want to go near any doctors, until Dr. Robertson came along. (Of course, it probably helped that Dr. Robertson's nurse gave Geordie candies every time he came to the office.) A couple of years later, we discovered Geordie's suspicion of doctors extended even to the doctor's wife.

We had been to Dr. Robertson's house for Christmas dinner and were surprised to find that, because the Mrs. Robertson came from the North of Scotland, she stuffed their turkey with oatmeal.

Geordie didn't like the oatmeal stuffing, but he knew enough to eat some for the sake of politeness.

Some time later, Dr. Robertson had to remove our son's inflamed appendix, and Mrs. Robertson visited him in hospital. Geordie looked at her suspiciously and said to her, "You know, Mrs. Robertson, I think it was your turkey stuffing that made my appendix bad." Luckily, when she told us the story, Mrs. Robertson laughed just as hard as the rest of us.

In spite of the war, the shortage of experienced men, and then my accident, all of us kept our sense of humour. Laughter helped us to survive.

I still laugh when I remember one man I'd trained for the farm. He became a damn good worker, who stayed with the job for quite a while. His actions sometimes puzzled me, especially one day when I saw him as I was heading for the barn. He was hopping on one leg, going to beat the band, heading for his car.

He noticed me and called out, "I'll be back in half an hour, Dave. I just broke my bloody leg!" With that, he climbed into his car and drove away.

I was dumbfounded. I went into the barn and spoke to his mate, who was working with the milking machines.

I said, "What are you doing letting him drive himself with a broken leg? How the hell did he break it?"

"Oh no, Dave" he said. "You've got it wrong. It's not his real leg. He's got an artificial leg, and that's the leg he broke."

With that, we both burst out laughing! I didn't know the fellow had an artificial leg—he was that good with it.

But there were many times I found myself very serious, hearing my father's wise voice inside my head . . . like the time four of my men were being criticized by Art Hay and one of his cronies. (Art Hay hadn't changed in the time we were in Hatzic.) The two of them were yelling at my men because twenty-one of the milk cows weren't producing as much milk as they had the year before. This year the cows were all heifers with their first calves, so the change in their milk production was puzzling—if you didn't know their history, which I did—and that history made me mad, I'll tell you!

I said, "It's not my men's fault the cows aren't producing. It's your fault. You gave the orders! Last year you made these men feed these heifers six times a day, so they'd produce more milk. You ordered their food measured into 'tip out' mangers. You ordered them milked every time they were fed, even through the night. You forced these cows to produce twenty thousand pounds of milk last year . . . far more than was good for the them! But you didn't care about the heifers! You wanted to brag about winning the provincial prize for milk production! You made a big mistake because you didn't know what my father knew. Years ago my father told me, 'cows are creatures of habit.' Art Hay, all last year you made my men milk these cows only when they were being fed! Now they're conditioned to let down their udders only at feeding time! But this year they're back in the large herd, where milking is no longer tied to feeding. Your prize heifers won't let their milk down so my men can't take it! It's not my men's fault that these cows are giving less milk. It's your fault! So, don't you blame my men!"

As I walked away in disgust from Art Hay and his crony, I could almost hear my father's voice saying, "Remember, Davy . . . cows are creatures of habit. Ye must be consistent with them tae be a good cowman."

My frustration had been growing ever since Art Hay became manager. Since his arrival, he had been busy proving—to me anyway—he didn't know his job. But Art wasn't my only worry. My father-in-law, Hector McMillan, who was now working in the calf barn, informed me I had an enemy in one of the workers, who was spreading lies about me. He was telling the superintendent, Pete Moore, and others, that I wasn't doing my job—but eventually Pete found this fellow out in his lies. The lies worried me but they did me no harm.

The big main barn with the arena burned down about 1946, not long after World War II ended. For me, this fire leaves an even bigger impression than the ending of the war. This huge building, maybe two hundred feet long, a hundred fifty feet high at the gable end, and about a hundred fifty feet wide, had stalls with over thirty-five head of horses—including Clydesdales—inside. There was also about nine hundred ton of straw and hay, with fifteen ton or more of barley and oats. Some of the workers still slept in the building, as I had when I'd first arrived at Colony Farm. Teamsters had rooms on the second storey, cattlemen and milkers roomed on the third. I'd just returned home from helping a cow calve in this building, on the night of the fire.

I was settling into bed beside Peggy—at about 11:00 P.M.— when the fire siren sounded. Up I jumped, pulling my clothes on as I ran back to the arena. The flames were already hot as Hades, yet some of the men ignored the heat and flame in their concern for the animals. They were running in and out, leading horses to safety. I immediately thought of the cow that had just given birth, and ran inside to help her.

When I got to her stall, I pulled the helpless calf over to the side away from the flames, then I grabbed the cow, trying to lead her, but she refused to move because she wouldn't leave her calf. By this time, the smoke was getting thicker. I could no longer hear the other men yelling, because of the roar of the flames and the screams of the terrified horses.

In desperation, I caught the calf by its hind leg and pulled it along the straw-covered floor towards the door. It wasn't old enough to stand, being not even an hour old and, with the damage the bull had done to my back, I couldn't carry it. Fortunately, the mother followed me and the calf into the outside air and safety.

Within no time, the inferno was so hot the firemen had to keep a hose on the roof of our house—three hundred feet away—to keep it from igniting. The arena burned to the ground as we all stood helpless, watching. It wasn't until later we learned the fire had killed one of our men, an older teamster sleeping inside. The poor fellow hadn't worked at Colony Farm very long, didn't know many men. In the panic, no one remembered him. About seven horses also died, in spite of everyone's frantic efforts to save them.

In the arena pharmacy area, I personally lost a book of recipes I had been collecting for years. They weren't recipes for food; they were for medicines used to help the animals: for iodine, iodex (an ointment mixed of Vaseline and iodine), and for liniments to be used on muscle strains. It was later I missed that book.

Pete Moore lost much more. All his photographs and ordinary breeding records of the Holsteins—kept since he'd arrived in 1910—were lost. This was probably the best breeding record in the world, and now it was gone. Gone too were ninety-seven volumes of American Holstein Herd books, the old Dutch Friesian Association books (which preceded the Holstein-Friesian Association of America), and also a complete set of Canadian herd books, some of them out of print.

It was lucky for us and for Colony Farm, I'd put all the breeding records since January into my office in the dairy barn—a separate building. Also, the breeding and identification records for the pigs were in the piggery. Because of this, Pete, Art Hay, and me were able to confirm the breeding certificates of all the animals on the farm . . . but the long history, the books, and all Pete's pictures, were lost.

Later, we learned the fire was the work of an arsonist, whose identity remained unknown for several years because he covered his tracks well. We were hoping the fact that someone died in the fire would scare him off, but he kept on. Maybe to throw suspicion off himself, next he lit a fire in the bunkhouse where he himself was rooming—though nobody knew that at the time. First he ignited the loft, then he lit one of the beds in a room with an absent owner. It was only the rifle ammunition—stored by the empty room's occupant—that prevented more deaths. The heat set off the ammunition and woke the confused boarders. They smelled smoke and started shouting. They reacted so quickly the fire couldn't get a real hold. The fact that he'd endangered the lives of twelve men was no deterrent to the arsonist, so the police warned all of us to be watchful.

"This guy is very dangerous," they said, but we knew that already.

By this time, the Mounties had a suspect, though they didn't tell us. They had contacted the police in Medicine Hat, Alberta, to see if their suspect had a record, but none could be found. Without more evidence, all they could do was warn us.

Before he was caught, this arsonist destroyed enough of the animal housing that Colony Farm ran out of room. By April, I had to transport eighty head of registered female Holsteins to a stock reduction sale, held in the livestock pavilion at Exhibition Park—sold because we had no place to put them, after that arsonist's work.

Between the bull's attack that left me wearing a steel back brace for years, the arsonist threat, plus Art Hay's annoying conniving, in 1947 my patience had just about run out with my job as herdsman. I decided to talk to Pete Moore about it. I said, "Pete, I'm fed up with some of the foolishness going on around here. I have to move on." I told him even Peggy was tired of the situation, and that sometimes she was mad at me because I was beginning to take my troubles home. Besides that, there were times when being a good herdsman conflicted with being a good husband and father.

I was responsible for about 230 milk cows, with many calving every year. These valuable cows were so well-fed and fat, they had trouble calving and I had to attend them more closely than I'd ever done before.

Some Saturdays, Peggy and I would have our good clothes on, ready to go into town shopping—the only day we were free to do it. Just as we'd be leaving, someone would come running, telling me there was a cow calving.

"Oh no!" Peggy would say. "Not again!"

Without saying much, I'd shrug, pull my coveralls over my good clothes and go to the barn. Jesus, Peggy would get up in the air about it!

Exasperated, she once told me, "Dave, you care more for those cows than you do for me!" I had no answer for her.

I was pleasing no one, it seemed to me. Eventually, I was just tired of it all and that's when I went to talk to Pete Moore.

"Just hold on a bit, Dave. Maybe I'll have something for you," said Pete. He didn't tell me what the "something" would be, but I trusted Pete. I bided my time a bit longer. One day when I was taking a look at a sick cow, Pete Moore's assistant showed up.

"Pete wants to see you in his office as soon as you can make it," he said. "Word is, he's going to offer you the job as farm manager of Tranquille." Then, as if I wouldn't know where it was, he said, "Up at Kamloops."

I smiled to myself, though I pretended to bristle as I said, "I know where Tranquille is. For all these years, haven't we been shipping our good beef to the biggest tuberculosis sanatorium in BC? And you think I don't know where it is?"

The fellow laughed and said "Just jokin,' Dave. No offence."

Later, when I talked to Pete, I didn't hesitate for a minute to accept the offer to manage Tranquille. In fact, as I was told I'd been accepted for the position, I swear I felt my father there—proud—as if he was standing right behind me.

Pete told me he'd suggested me for the position and spoke on my behalf. "I told them I know you can handle the troubles they're having up there," he said. When I asked what he meant by "troubles," he wasn't too specific. He said only, "We're spending too much money up there."

Though I'm sure at that time Pete thought he knew everything there was to know, he might not have realized the extent of the double dealing and graft that was causing the troubles. He was now Supervisor of Farms under the Provincial Secretary, including Colony Farm, Tranquille, and Colquitz on Vancouver Island.[1] He had his hands more than full.

To me he said, " I know you'll not be intimidated by anyone, Dave. You'll do what you have to do to get that place back on track."

Though I was pleased by his vote of confidence, I wondered . . . just what kind of a situation am I gettin' myself into?

CHAPTER THIRTEEN

Tranquille

Dear Smith, the sleest, paukie thief,
That e'er attempted stealth or rief . . .
—Robert Burns, "To James Smith"

The day Peggy and I arrived, the weather couldn't have been more beautiful. Peggy was very impressed with the setting—fertile bench land on the north side of Kamloops Lake. Tranquille was BC's largest sanatorium in the 1940s.[1] Pete Moore met us and introduced us to the doctors and the rest of the staff. We learned they had already fired the man whose job I was taking, giving him two months' severance pay. I was told only that he'd been helping himself to too much, though for the most part, the people who worked here were described as honest.

As we toured the large grounds, we could see the hospital and its support buildings functioned almost as a little town would function. Each building provided some needed service to the people who lived there. The Greaves Building was the most modern and had an operating room on the top, the fourth floor. Men and women patients were assigned separate living quarters, with the women in the Main Building and the men in the Infirmary when they were bedridden. Later, when they were up and about, male patients were moved into the pavilions and the women were relocated to small cottages. The nurses had their own comfortable residence, close to their work. The families of medical doctors, farm staff, and gardeners lived in private homes set back some from the hospital buildings. We were all free to use the hospital store and post office located in the centre of things. The store was

Note to chapter 13 is on page 241.

necessary because few people working here had cars to take them into town to shop. About a year later, Tranquille became part of a bus route, so the store and its groceries were phased out. On the day we arrived, I remember Peggy was especially impressed with the beauty and the colour of the grounds. The green grasses, crops, and coloured flowers stood out bright against the soft brown sands and pale sagebrush in the surrounding countryside.

Unlike Colony Farm, we were told the vegetable gardens here had no patient labour used in them, although patients did do lighter volunteer work in the wards. But the lack of patient labour wasn't the only thing that was different from Colony Farm. I wasn't there long until the problems Pete Moore warned me about—plus a few new ones—became obvious.

Pete had said, "You'll have an uphill time, Dave, but I know you can take it." And he was right. It wasn't easy at first, because I wasn't quite sure who to trust. I also knew I had to be very careful in dealing with the unionized workers, who were organized and a force to be reckoned with.

The running of Tranquille was costing about $600,000 more than the government appropriated for it, and there were many reasons for that. For example, Pete Moore told me there were far more people working there than should have been. In spite of the money spent, productivity was low. In the Kamloops climate, I knew Tranquille should have been producing three crops of hay a year, sometimes four, but often they couldn't even get two crops. I couldn't believe that a farm as big as Tranquille had to buy hay to feed their cattle.

For the first three months I observed the hay production process, then I laid off twelve of the men involved. This got the attention of the rest of the workers and most of them knew I meant business.

After that, we started cutting the hay when it should have been cut. We'd begin by cutting mostly alfalfa, about the twelfth of May. With that early start we harvested three crops—more than enough to feed our cattle. From then on, there was no more buying hay. Tranquille's operating costs started to come down.

Next, I tackled the problems with the irrigation system. There was plenty of water available, but half the time the men weren't getting the bloody stuff to the fields. Instead of checking to make sure the water was directed and flowing properly, they'd be sitting around together telling jokes. One of the irrigators was always sitting on the fence when I drove up. I warned him he'd be fired if he continued. But he and a few other men still didn't think I would do it. They soon found out different. After I let one or two of them go, the idea began to dawn that things were changing at Tranquille. They discovered that when I could prove I had good reason to fire someone, their union couldn't stop me. They were now required to work for their $65.00 a month, plus room and board.

I was amazed to discover the piggery was also in a sorry state. The pigs were dying at a rate of about seventy a month. Why? Because they'd been running the pigs outside, on the same ground for years, and the ground was contaminated. To get rid of the problem, I had to convince the superintendent of the need to build a piggery with a cement floor. With the pigs running on cement, we could wash it down every day and keep the animals safe from disease. A piggery had the additional advantage that fewer men were needed to care for the pigs, because they were all kept within one enclosure. With more pigs making it to maturity and fewer men needed for their production, we paid off the cost of the piggery in no time.

But while I was cutting costs in farming, there was no cutting corners when it came to the patients. Dr. Stocker, the Medical Superintendent, made it very clear. To me he said, "Use nothing but the best food for the patients." He soon proved he meant what he said.

I went to him after receiving a shipment of thirty cows from Art Hay, sent to be used as food for the patients. The problem was, I discovered these cows had been dry with mastitus, so I went to Dr. Stocker and explained the situation. Dr. Stocker contacted Art Hay and immediately I was told to return the shipment. Later, I heard the meat was sold in Vancouver, for use on the ships.

Dr. Stocker said, "Art Hay should have known better. We have always used only the best meat for the patients."

Then I took a close look at the cookhouse. It was amazing what I learned just by eating some meals with the men, instead of going home to eat with Peggy and the boys. One man had been eating in our cookhouse for five years, but he wasn't even working at Tranquille. Another man had the title "Stable Boss," but cared for only six horses. He spent his time mostly in the town beer parlour, so whenever I went to look for him in the stable, he wasn't there. I only met him at the dinner table, while he was filling his face with Tranquille food. He was another one who thought his job was safe because he belonged to the union.

"You can't fire me," he said, but he was wrong.

Not that the union took my actions lying down. They had meetings about the firings and tried to put pressure on me to force me back to the old ways, but they didn't succeed. As long as I had proof that the firings weren't vindictive and were based on good reasons, there was nothing the union could do.

As farm manager of Tranquille, I had to learn some things my father never heard of on the Scottish farms he managed. One of the most annoying lessons for me, was the paperwork required to get the funds for running the farm. All the money came from yearly government appropriations. The appropriations were taken to Victoria in November, so that the legislature could approve our budget by March—including wages, new machinery, and any other expenses. Every amount requested had to have an itemized list attached, to justify the expenditures.

Though Pete Moore and I worked together to develop the budget, I had to provide the detailed lists. I had the responsibility of making sure every need was covered, including allowing for possible unforeseen emergencies that might arise during the following year. This was a big responsibility and it took time to plan ahead carefully. It was almost impossible to get money to cover unforeseen events. This budgetary process was one of the reasons I had to work so hard to get permission to hire Japanese help. I needed the workers for the market garden.

The market garden also had not been producing as well as it should, because its fifteen acres weren't being maintained. None of the White workers wanted to do the tedious hand-weeding and hoeing. I could understand how they felt, because I didn't like weeding, myself. When the men did work, they worked half-heartedly. The crops were suffering, so I knew I had to solve the problem quickly.

When I was down near Kamloops, a well-tended hop garden made the answer to our market garden problem clear to me. I stopped and began talking to the foreman at the edge of the plot. I mentioned I'd noticed the hop garden's Japanese field hands were working very hard. The foreman told me some of them were soon to be laid off.

On impulse I told him, "Send them up to Tranquille and I'll hire them as casual labour for fifty-five cents an hour."

Though I wasn't offering much money, I knew I was going out on a limb to offer them anything at all. There were two reasons for this. First, it was only a couple of years after the war. Many Whites still hated the Japanese, thinking of them as "the enemy." Second, there was no money in Tranquille's government appropriation to pay them, and I wasn't sure how to get any.

A few weeks later, Pete Moore came through for me. He stopped in at the end of a long weekend fishing trip and Peggy invited him to stay for supper. First, I told him about hiring the Japanese. Then I took him to the now healthy and orderly green fields of the market garden, so he could see for himself how much work the Japanese had accomplished in just a short time. Pete was impressed. They had hand-weeded and hoed every field and were in the process of harvesting some of the vegetables.

Pete said, "I've never before seen so much work done so fast. You did the right thing to hire them, Dave. For the time being, pay them out of petty cash, and I'll authorize it."

The next time he went to Victoria, Pete got an amount properly set aside in our budget, so Tranquille could continue to benefit from Japanese fieldworker assistance. Once we had the gardens all cleared up, I needed only four Japanese women to maintain the whole fifteen acres. They outperformed a whole crew of White men. They were that good.

But their work didn't please everybody. People soon let me know they were very upset, accusing me of taking jobs out of the hands of White men. And they didn't only complain to me. No sir. I know of at least one woman who, after she got nowhere threatening me personally, wrote to the government in Victoria demanding I be fired, along with the Japanese help.

Of course, it meant nothing to the complainers that these people were Japanese Canadians who had been moved to the interior from the coast during the war. Nor did it matter that the white workers didn't want the jobs. The Whites just didn't trust the Japanese, nor did they want them in the labour market. Of course, I didn't let the complainers have their way. I knew good labour when I saw it.

My frustration with all the paperwork at Tranquille was eased by a phenomenal woman, my secretary, Geraldine Rossitch. (Everyone called her Gerry.) She was the only secretary I ever had all to myself for farm business. She did the Tranquille payroll and helped me with any correspondence I had to produce. She was honest, determined, and loyal, and much like Peggy in that she wouldn't let anyone say a word against me.

One day, I happened to be just outside the office door, when I overheard one of the farmhands complaining about me to her.

"That Caldow's always on my back," he said.

Then I heard Gerry. "Are you doing your job? If you're doing your job, Dave won't bother you," she said firmly. She took no nonsense from anybody.

She'd had a tough life, coming from a family of fourteen, and she faced some prejudice—on the part of a few people—because she was staunch Roman Catholic. She got her stenographer's training in a convent. She was so shy, she would blush if the office conversation ever turned to the breeding of pigs and cattle, but Pete Moore certainly broke her into the language of the outside world, with his swearing.

One day, after I got to know her better, I teased her. I said, "Christ Gerry, ye'll never get yoursel' a man. You still sound like you're married to the church!" But she just laughed. She'd always been independent.

She placed a lot of stock in religion, but that was her. It didn't bother me that she was a Catholic. I knew the Protestants in Scotland were no bloody angels. They were the ones who turned people out in the cold Highlands, burning their houses so they couldn't return—and the Protestant ministers helped them! They were a bad lot, so I didn't see any reason to pick on Gerry Rossitch because she was Catholic.

Gerry was trustworthy. That's more than we would have said about Colony Farm's head secretary, and the one under her in the Colony office, at that time. Pete, Bruce Richardson, and I knew we could talk about any business in front of Gerry, and it would go no further.

One day, Gerry asked me if she could be late for some reason or other. I told her, "I trust you, Gerry. All I care about is, can you get the job done—and you always do—so as far as I'm concerned, you can come and go when you please. You don't have to ask me." That's how I felt, and she never let me down. She was the best secretary I ever had.

Later, Pete Moore told me the details of the story about how the farm manager at Tranquille before me had been fired. It seems at first, the medical superintendent had suspected only one old butcher of helping himself. He knew the old guy slaughtered three beef and sixteen pigs every week (all owned and raised in Tranquille, o' course). This beef, hams, and bacon was supposed to go to feed the patients and staff at the sanatorium, but that wasn't quite what was happening. The old butcher was keeping back a quarter of beef a week, plus some hams and bacon, then selling them cheap to a Chinaman who owned a grocery store in Kamloops. Once a week, the Chinaman would drive his truck right up to the slaughterhouse at Tranquille and load his booty, bold as you please. The butcher was pocketing the money from the sale of the meat.

Somehow, the medical superintendent—officially in charge of the farm and hospital—got wind of what the old guy was doing.

He devised a plan to catch him in the act of selling the meat, but the plan caught more than he bargained for, and almost got someone murdered, to boot.

The medical superintendent waited until an official from Victoria was at Tranquille on business. Telling the official what he suspected, he suggested they get the farm manager and the three of them would go over to the slaughterhouse that night, trying to catch the old butcher in the act of selling meat to the Chinaman. They found the Chinaman's truck parked, just as they had been told it would be. The medical superintendent and the other two men listened at the door of the slaughterhouse. When they judged the moment right, the plan was for all three men to burst through the door and surprise the two thieves.

As planned, the medical superintendent and the official from Victoria banged the door open, but only the two of them entered quickly, yelling, "What the hell is going on here?" The Chinaman and the butcher froze for a moment. Then—eyes wild as hell— the Chinaman grabbed a bloody cleaver and started swinging it as he roared towards the superintendent and the official. They froze in their tracks.

Lucky for the two of them, the butcher was only a thief, not a murderer. Reacting quickly, the butcher grabbed the Chinaman's arm from behind. The cleaver fell to the floor, saving at least one life.

It was lucky for the old butcher that he acted. If he hadn't, he'd have not only lost his job, he could have been up for murder. Pete told me the medical superintendent telephoned the police right from the slaughterhouse and the police took the two criminals away, without further trouble.

After the excitement was over, the medical superintendent began to wonder why the farm manager had hung back, not entering the barn with himself and the Victoria official, as they had agreed. A check was made of the duplicate grocery bills at the hospital store. The bills showed the farm manager never bought any meat for himself, for use at his home. He must have uncovered the old butcher's dishonesty, but instead of reporting him to the medical superintendent, this farm manager decided to take a weekly cut of the stolen meat for himself, and—who knows— maybe a cut of the stolen profits, as well.

Though the medical superintendent knew the farm manager before me was guilty of using his position for his own gain, proving him guilty would have been difficult. His cohorts were now gone. Instead of pressing charges, it would be simpler to fire him. This action had the added advantage of keeping Tranquille's problems out of the public eye. That's why even I had to wait to discover what happened.

On one of his visits, Pete also told me the story of what happened to the arsonist, back at Colony Farm. I didn't realize it, but the fella'd been working at Tranquille for awhile, maybe trying to get the heat off himself. Then he returned to Colony Farm and started making phone calls, threatening the police with more fires. So he wouldn't leave fingerprints on the phones, he'd used a pencil to do his dialling . . . but sometimes he dialled with the lead point down. This was his undoing, because he'd also made some notes in his room with the same pencil. Somehow, the police matched the traces of lead on the phones to the pencil and notes in his room. When the police told him about the match and the fact they'd been tapping his phone, I guess the arsonist figured they had him, for sure.

Pete said the police detective told the arsonist, "There's no use in carrying on like this. We know everything you burned, bridges and all. You might as well come clean."

With that, Pete said the arsonist broke down and confessed to everything. Everyone was relieved to know he was going to trial.

After we got the arson at Colony Farm and the corruption at Tranquille cleared up, I enjoyed not only my work, but also my off hours. Peggy enjoyed Tranquille, too. She was well-liked by the doctor's wives and other people in residence, so we had a rich social life. Peggy began volunteer work with the patients, as Geordie grew more independent. There was plenty for her to do. Sometimes she pushed library carts filled with books to the patients who were bedridden, or helped the occupational therapists teaching the leather work, knitting, and other crafts, helping the patients pass the two or three years it sometimes took for the long rest treatment to achieve recovery. Peggy helped mostly in the womens' wards, because male and female patients were strictly segregated, not even allowed to eat together in the dining hall, though I know for a fact some couples managed to get together. I even heard of one or two marriages that began with chance meetings in the cafeteria line-up for food. (Of course, Peggy had to go through a training to learn how to protect herself from getting TB.)

Visiting Dobbin at Bob's Silver Creek Ranch.

It was about this time we began our family tradition of going up the Hope-Princeton Highway to visit my brother Bob and his wife Mary, at Kelowna. We'd stay only a short time, but Geordie would remain all summer. This was good for him because they had three boys and he benefited by their company. Coming back, we usually would tour through Osoyoos on the way, then cross the border and stop at Winch—because Peggy liked to shop for clothes and linens in the States—then we'd drive through Stevens Pass, on the way home. (Peggy was a thrifty shopper, yet she insisted on getting quality for her money. I admired that in her.)

It was one of our pleasures to find a shady spot to spread a blanket on the grass. Then we would sit together eating from a box of cantaloupes purchased at a roadside fruit stand. I always took a knife and spoons along on these trips, for this purpose.

In these days, fruit was only available in season, so each year we'd have to go without cantaloupe for maybe eight months. We could hardly wait for the new crop to be ripe and ready. I can still smell the sweet juiciness as I cut into them. We weren't even fussy about scraping out the seeds, the way we would be at home. Sometimes I would eat about a dozen cantaloupes at one go, while Peggy would eat about two. I know memories are always better, but it seems today there's no cantaloupe as delicious as the fruit we shared on those summer journeys.

By the end of 1953, I was working my seventh year at Tranquille among rumours that the government would soon be closing the place. Medical advances were the prime reason. By this time, doctors had their choice of three tuberculosis medications that would cure the infections, allowing their patients to remain at their homes and work instead of being isolated in hospital. Though this was good news for the patients, it didn't appear to be good news for me and my family. I began to worry what would happen to my job. Peggy worried even more than I did. One more cause for concern was the fact that Pete Moore was thinking about retirement.

On one of his visits, he'd shown me a letter in which the government requested him to consider staying on for another couple of years past retirement age. They said his knowledge was needed to help them through the difficult times of coping with the damage caused by the arsonist's fires and the lack of funds. Pete made the mistake of showing the offer to Art Hay, before he showed it to me.

We found out Art had reacted true to form, as I knew him. Art had gone to Victoria with a couple of cronies—not to help Pete, but to speak against the extension of his appointment. As a result, the government withdrew their offer to Pete and appointed Art his successor.

Art Hay got what he wanted, but it didn't last long. He was gone before I had to worry about him, because his performance proved he couldn't do the job. Bruce Richardson was appointed Superintendent of Colony Farm after Art left.

Pete Moore was given a big retirement party at Colony Farm. Then we organized another for him, at Tranquille when he came up for the last time. Good as he was at his job, Pete was no good at

I drove Lt. Gov. Wallace on a tour of Tranquille. He said, "I'll be happy to report this is a working farm, not a showplace." (I'm the driver.)

saying goodbye. He had tears in his eyes and he choked when he tried to give his speech. He and I remained friends for years after.

One day, I mentioned my concerns about my job security when I was talking to Bruce Richardson. He said, "Dave, Bob Gardner's now sixty-five and has to retire from Colony Farm. I'd like to see you in his job. Why don't you apply for it?"

I took Bruce's advice and sent in the application, then I went for an interview with the government representative. The fellow was very pleasant.

He said, "Well Mr. Caldow, I won't need to ask you any questions about your ability to do the job. You've more than proved yourself at Tranquille." Soon, a letter of acceptance arrived, appointing me manager of Colony Farm.

My success was greeted at home with mixed feelings. Peggy liked her Tranquille friends so much, she didn't want to leave. By this time Geordie was about ten years old, old enough also to make his voice heard. He told me in no uncertain terms, he didn't want to leave, either. But with Tranquille rumoured to be closing, their protests were to no avail. Leave we did, for an even bigger challenge.

CHAPTER FOURTEEN

Colony Farm—The Final Years

In Ayr, Wag-wits nae mair can have a handle
To mouth "A Citizen" a term o' scandal . . .
                    —Robert Burns, "The Brigs of Ayr"

In 1954, we began the new year at Colony Farm at Essondale—as the mental hospital was now commonly known. Both Peggy and Geordie settled in with few problems because returning to the same wee house—built for us when I was herdsman—had the feeling of returning home. However, my new job was not so comfortable.

Before my first official day of work, Bruce Richardson stopped by the house to say, "Dave, I've some bad news, though it's not going to change my mind about you. Some of the workers have circulated a petition protesting your return. Some of them have already quit."

I thought to myself, *They must have heard I laid off the slackers at Tranquille. I suppose they panicked. The worst of them must have decided to quit while they were ahead.*

To Bruce I said, "If they've left, it's good riddance. They must know they're not reliable, and if that's so, I don't want them working with animals and people who need them. Don't worry Bruce, I'll talk to those that're left, let 'em know where I stand before I begin."

From this moment, I knew I'd a big job ahead of me, though I didn't realize just how complicated. In appearance, Colony Farm seemed much the same, although major changes had occurred in the hospital. Since the end of World War II, the medical staff had

Notes to chapter 14 are on page 241.

been working toward separating the mentally ill from the mentally deficient and senile patients, both in terms of their housing and in the way they were treated.[1]

That damned arsonist had created some major problems. Patient population numbers had increased. The burned buildings were sorely missed and took time to replace.[2] Some—such as "The Cottage" which once had housed only patient labourers— had to be renovated to house a few extra patients who needed more treatment before they could work.[3] The Crease Clinic of Psychological Medicine—a large building named after Dr. Al Crease—had been opened on the hospital grounds November 16, 1949, just one year after I left for Tranquille. It was a large four-storey building, including a lab, medical and surgical offices, a library, and room for 325 patients. Facilities for tennis, lawn bowling, and ping pong were included in the planning.

Colony Farm with Mary Hill in the background.

The Clinical Psychological Medicine Act, which came into effect January 1, 1951, now permitted some patients to be admitted to hospital on the recommendation of two doctors, with a consent form signed by a relative.[4] Some were still committed by the courts, and some were voluntary patients. Before this, only the courts had the power to commit the mentally ill, so permission to be discharged was hard come by. However, after 1951, some improved patients could be approved to go home—out of Crease Clinic in four months if they made enough progress. So patient care had changed somewhat, but most of the jobs we did to produce patient food at Colony Farm had not changed much, except for the incompetence of some of the staff.

The herdsman was an American who had worked in farming on the prairies. He should have had a relatively easy time of it, because he was herdsman in a state-of-the-art facility—Pete Moore's "milking parlour and loafing barn system" had been functioning since 1948 or '49.[5] It was one of the finest dairy operations in the world.

The cows lined up—eight at a time—outside the milking parlour, where "electric eyes" opened and closed the doors for them.[6] The cows walked through an antiseptic foot bath, then proceeded to the milkers and milking machines inside. (It was a new idea to bring the cows to the milkers, instead of the other way around.) The milk was pumped from the cows directly into glass containers, weighed automatically, then moved to the pasteurization unit. Everything was clean. The floor of the milking parlour was an iron grating, allowing dropped manure to be continually flushed away by a stream of water. After milking, the cows returned to their "loafing parlours" . . . three 160 by 40 feet, open-floored, all steel, quonset-type barns with no stanchions or stalls. Hay was always available filling long wood and concrete mangers, placed along the centre of the concrete floors. Deep wood shavings—renewed every day—provided softness underfoot. When not being milked, the cows could sleep, eat, and move at will. Pete figured his milking parlour/loafing barn system would result in a 20 percent faster milking time and a 50 percent reduction in labour. It was a good system, needing reliable men to make it work, but the herdsman was not reliable.

He was a conniver, a sneaky bugger who wanted to get rid of Bruce Richardson because he knew Bruce wouldn't approve of his antics. This herdsman was in cahoots with some managers in

Essondale, who also wanted Bruce Richardson out. To them, Bruce was too honest. It took a while for me to understand their collusion.

I knew Bruce was a capable man in his work, but he wasn't a good judge of men. If he had been, he wouldn't have hired the American herdsman. What kind of herdsman neglects proper supervision of the milking?

I discovered some of the cows had mastitus because the dairymen weren't milking them right. The herdsman had done nothing about it. I had to order many of the cows to the slaughterhouse, the infection was so bad. With the cows that could be treated, it took hard work to clear up the persistent disease.

"Get started right away," I said. "Get those men milking on time and properly. Reduce the protein in the feed and inject sulphanulamide into the udders. That'll cure 'em, but we'll have to keep at it." These was standard procedures—I shouldn't have had to tell him. A good herdsman would have known, but I figured I'd give him a chance to do the job right.

Next, I found he'd ruined our very successful breeding program. He'd discontinued artificial insemination, the method I'd left in place when I went to Tranquille. He was letting the animals breed at night, putting them in fields together. This method produced calves, but it was impossible to closely supervise the breeding, so we weren't always sure which bull mated with which cow. This knowledge was crucial to the credibility of Colony Farm's breeding program, and to the profits to be made from it.

Christ, I thought, *this guy has no idea what his men are doing with the dairy stock!*

I told the American in no uncertain terms, I wasn't pleased with his actions. Then one sunny winter morning he made me really angry. I found him letting valuable cows outside on frosty ground. The cows' legs were shooting out sideways, slipping and sliding on ice!

"I can't believe my eyes," I said. "And you call yersel' a herdsman! What herdsman would take such chances with the cows' safety? What the hell are ye doin' man! Get those cows back inside!"

The American yelled back, "I'll do no such thing! You mind your own goddamn business! I'm doing my job as I see fit!"

"You're doing this on your own responsibility," I said. "If there's any of these cows hurt, you'll pay for them!"

With a sly smile and calm as anything he replied, "I don't know how you'll do that, Caldow."

Gritting my teeth, I said, " Just let an injury happen and you'll find out!"

The day after I challenged him, he came into my office wanting me to sign a Workmen's Compensation form to excuse him from work. Before I'd even read it, I knew it wasn't the work he wanted to escape. He wanted to get away from me!

"Ye've got down here ye hurt your back lifting a big door during that storm, last week," I said. "Do ye think I don't know you weren't even here when the storm was on? Don't you remember, you were in town for two days, then?" The herdsman went red-faced, but I thought it strange his sly smile returned as he turned away to walk out the door.

A few days later, Bruce Richardson informed me, "The herdsman's now off. He's on Workmen's Compensation."

I thought this news was pretty mysterious since I hadn't signed the Workmen's Compensation form.

"Did you sign his form?" I asked.

"No," said Bruce, "Didn't you?"

I filled Bruce in on some of my discoveries and suspicions. Later, I learned it was the business manager—one of the herdsman's cohorts in Essondale—who had signed, probably in return for a few gifts of stolen vegetables, milk, and cream.

Bruce spotted the herdsman back at the farm a few evenings later, talking with one of the cooks at the side of an outbuilding. Staying out of sight in the shadows, Bruce followed them. Behind a building, he watched as they divided cream and vegetables they'd stolen from one of the kitchens. They spotted Bruce and ran, taking the evidence of theft with them.

Not long after this, the herdsman quit of his own accord. I guess—between Bruce and me—it was too hot for him. Bruce saw to it the cook was kept under observation, then fired when he was again caught in the act of stealing.

One warm night after I'd finished work, Peggy, myself, and Jeannie Cruikshank (my sister Jess's daughter) was outside sitting on lawn chairs, talking, enjoying each other's company. (Jeannie lived with us for a time, when she first immigrated to Canada.) We were relaxed, surrounded by the calls of night birds, the songs of crickets, and the sounds of farm animals. It was a dark night with no moon, but peaceful—or so it seemed.

Peggy said, "Dave, would ye go get us a half dozen Coke from that new pop machine ye've had put in?"

"Sure," I said, rising and ambling down the path to the barn. I was thinking it was a good thing I'd agreed to have the Coke machine installed outside the barn. Put there mainly for the convenience of the patient workers, it was handy for us, too. Then I remembered, *In order to get more than one pop at a time, I'll have to go inside the barn and get myself a tool to open the machine.* As I walked down the path in the dusk, I heard only the crunch of my own feet on the gravel. I'd never a thought of any danger.

The heavy barn door groaned as I pushed it aside. As I moved through it, into the deep darkness inside, I groped for the light switch. Suddenly I heard the swish of something moving quickly through the air. Before I could react, my wrist exploded with pain.

"Jesus," I said as I hunched over and grabbed for my arm. Then I heard footsteps running—first clomping on the barn floorboards, then crunching on the gravel—past me and away.

I spun 'round quickly and caught a glimpse of a man turning the corner of the barn. My first thought was that the American herdsman had returned, but then I saw enough of this man to recognize the cut of his coveralls and his unusual manner of running. In that instant, I remembered. This thief was one of the people sent to Essondale for job training from the Prince George Welfare office. I knew I'd pegged him dead to rights.

Half in shock from the pain, I ran back to our house, giving Peggy and Jean a fright when they saw my blood-soaked sleeve and hands. As they looked at the wound, I noticed my watch was missing. When I found the watch later—back at the barn—my leather wrist strap had been cut through. I'd been hit by a steel carpenter's square which was lying nearby, where the thief had dropped it.

Peggy was right when she said, "Lucky he didn't connect with your head, Dave. He'd have split it open with the force of that blow."

"I swear t' ye Peggy," I said, "if I ever see him again I'll hit him on his head with my walking stick!" (I was still using one because of my back injury.) "I'll knock him cold, the thieving coward!"

When I'd surprised the thief, he'd been stealing tools, as I could see by the empty outlines on the wall where they hung. I realized this fella knew where the tool room was located, because he'd been working as an orderly at Essondale long enough to find out. After this, I knew no amount of honest work would reform him. He'd made his choice and all he wanted to do was steal.

I called the police and when they investigated, they said it was obvious this fella was a professional. He'd used a diamond drill to cut a hole in the window of the tool room door, then put his hand through to open it.

I never did see the thief again, never got the chance to beat him with my stick. He'd run off and—from the responses I got to my inquiries—I don't think the police were too interested in pursuing him. I suppose a blow to the arm and a few stolen tools don't make a thief top priority with the authorities.

At this time at Colony Farm, some patients still enjoyed the privilege of working with the animals and also in the fields. About a hundred and seventy of these patients were still housed in "The Cottage," where the farm cookhouse was also located.

Their labour was used in many ways. Patients waited table for the farm workers, before they ate in their own supervised patient dining area. There were about five gangs of patient workmen who actually did farm work, with ten men in each gang. There were also about twelve or fifteen patients "on parole" who could be trusted to do farm work without supervision, because they wouldn't run off. These men often worked in the cow barn, or with other animals. But the use of patient labour changed for the worse, once the politicians got into the act.

It happened in the early 1960s, about seven years after I'd returned. The government decided to change the policy on patient labour. Two CCF legislative members, a father and son—as I remember—set themselves up as critics. It seemed one or the other of them was up at Colony Farm all the time, watching the patients work. Each time the politicians came, they'd be giving stories to the paper about how unfair it was to exploit patients for the benefit of the farm.

Bloody politicians! All they cared about was theirselves and getting elected. They didn't have the sense nor the concern to see that Dr. Doherty, the man who originated the patient work policy, was right—the outside work helped the patients to get well! Anybody at Colony Farm who had worked year after year with the patient labourers could have told them that, but there was no way they wanted to hear it. The government had a cost-cutting agenda and there was no doubt that using patient labour was expensive, due to the needed supervision. During the war, the shortage of men to do farm work also lowered the farm's profit-making ability. Without enough manpower, the farm struggled to produce enough for the patients to eat, with little left over to sell for profit. After the war, the increased number of patients used more farm food, also leaving less to sell for profit. So the politicians claimed they needed to save money.

Then the unions backed the politicians. They claimed using patient labour robbed their members of jobs. Now, that was a bloody crazy statement if I ever heard one. For almost every working patient we had to hire a supervisor, so how could that be reducing the number of jobs available? But unions like to sell their members on how useful they are at protecting their rights, so the unions wouldn't hear anything from us to prove them wrong. At this time, I don't know why I wasn't fired, because I sure told people what I thought of their damn politicking antics.

I knew you don't take humans and lock them up, if you can help it. Years ago in Scotland, my father taught me, you don't even do that to a horse. "A confined animal learns only bad habits that you can't break them of later," he said. It was common sense.

I knew there were many ways confined humans reacted just like horses. All animals need to be encouraged to learn useful skills. But nobody would listen to me, or to any of us who worked in the fields at the farm.

Based on past experiences with the poor bloody patients, I could have predicted what the future would hold. After the hospital patients were barred from the fields, they were locked in rooms most of the time. Oh sure, they'd be taken out for physical therapy sessions, or to make handicrafts, but that's bloody nonsense for good, able-bodied men.

When the confinement drove them wild—the way they used to be when waiting to get out in the fields during four or five days of rain—the only thing the hospital staff could do was drug them to keep them quiet. I saw proof of the drugging later when I ran into former patient labourers outside in the sun, walking between buildings. The poor souls' faces were blue—which I later found out was a side-effect from one of the anti-psychotic drugs, chlorpromazine. I felt sorry for the poor buggers, victims of stupid politics.

But I knew there was no point agonizing over a situation I had no power to prevent. I had to get on with my work, which was more than enough to keep me busy. There was not one unionized civil servant who worked as hard as I did as farm manager—none worked even near as hard.

Besides managing the farm, I had volunteered for the job of sheepman because they couldn't find anyone else to do it, at the time. I'd had to convince Bruce Richardson I could handle both jobs, but that wasn't hard to do. At the start we had only fifty ewes, so—when they weren't lambing—the work was only an extra hour a day. Working with the sheep was a great pleasure for me, taking my mind back to Scotland, to my father, and the work on my uncle's farm.

Maybe my love for the work was part of the reason why Colony Farm sheep were so successful at winning prizes later on. I remember one time when I was showing Horned Dorsets at the Pacific National Exhibition—rare sheep at the time in BC—we took all the prizes. First, second, and third in each class went to Colony Farm. Yet we received none of the prize money because we were a government institution. Instead, the only man showing against us was awarded the money. In spite of that, he still complained to the judges and wrote a letter to the Provincial Secretary in Victoria saying we shouldn't be allowed to compete in the PNE. His reason? As an individual, he didn't have enough money to compete with us.

He had no idea how hard it was to get money for anything at Colony Farm. We still had to submit for money in September, then it would come up for discussion in March in the legislature. Then the opposition party would always come out against any request we made for finances. By the time the legislature finished with its arguing, it could be months later. Furthermore, we never knew if we'd get some or all of the needed money, until this whole process was finished. Any other competitor at the PNE had an easier time getting money than did Colony Farm.

I loved working with the sheep and enjoyed being with my family, but when the sheep lambed, my family got short shrift of my time. Peggy wasn't very happy when I had to be checking the pregnant ewes all the time. As time went on and I got busier, this became more of a problem. I spent more time at my job as the years passed.

As a result, I don't believe I was a very good father. I don't think I was cut out for it—yet I tried. When Geordie was in junior high school, Peggy and I both took the time to go to the Parent Teacher Association meetings. We helped the PTA buy milk for the Indian children, one of its projects.

When Geordie became very involved in sports, I enjoyed going to most of his games and, as any parent knows, those games take time. Geordie played baseball, softball, basketball, soccer, lacrosse, and he competed in track and field. He could toss and catch a ball with either hand, but he didn't get that from me. It must have come from my left-handed sister Jess, or my uncle Jake, who could shear with both hands. Geordie's lacrosse games were the best to watch, as far as I was concerned. I often drove eight or ten of his teammates to games and practices in our truck.

Geordie was a good lacrosse player because he was very fast. He could have been a world-class runner, if he had been more dedicated. Just how close he came was obvious when he ran in the Newton relays, in Surrey in 1959. He was running the hundred-yard dash in competition with Harry Jerome, who became BC's all-time most successful Olympic runner a short time later. In this race, Harry Jerome was the first high-school runner in BC to run the hundred-yard dash in under ten seconds. Harry Jerome ran it in 9.9 seconds. Geordie came second, with a time of 10.05.

Geordie was fast, no doubt about it. I'll give him his due, he never waited on a bus to go to school. He'd often run the whole way, which was from Colony Farm to Port Moody, about six or seven miles. Or he'd run, then thumb the rest. But as a young fella, he never took his running or anything else seriously. Whatever he did always had to be a lot of fun for him, or he was done with it.

In some ways, I blame Peggy for that. Christ, she adored Geordie. She couldn't see anything bad about him, even when he got up a ways in age . . . and there was times when I could have killed him.

Like the time I came home tired after a day's work. Peggy met me at the door and the minute I saw her I knew—from the look on her face—something was wrong. I thought immediately of Geordie, who was now in his second year of university. He'd had his chauffeur's licence since he was seventeen, and—to help pay for his schooling—in summers he'd been working driving freight trucks and buses from Port Hardy over logging roads to the air force base at Holberg. I knew anything could happen on them logging roads.

"All right, Peggy," I said. "Out with it. What's happened? Is it Geordie?"

"He's been in a fight, Dave—but ye'll not be mad at him, will you? He needs your help." With this she handed me a letter addressed to George. She said, "I only found out because I opened it by mistake."

As I was reading the letter, I became furious. The letter was from a lawyer, hired by a man who claimed Geordie had broken his arm, ribs, and nose in a fight. The man was suing Geordie— meaning us, of course—for nineteen hundred dollars, a goodly sum in those days. This was about 1961.

I said to Peggy, "What's the matter with our son? Doesn't he know we've no money to pay for this?"

"Please, Dave," said Peggy. "Calm down. I want you to go up to Port Hardy and talk to Geordie. That's the only way ye'll get to the bottom of it."

At first I said, "Are ye daft, woman? I'm not paying for any plane ride up there. How can we afford that and the mess of costs this lawyer's trying to stick us with?"

But Christ! Peggy would have nothing but that I had to get on the plane and go up there right away. As I often did with Peggy, I gave in, and up I went.

After I arrived, I was glad I'd taken Peggy's advice. Geordie's story made me view the whole event differently. It appeared he and some of his friends had been at a steel-band dance, at a hall at forty-first and Fraser in Vancouver. As Geordie was about to leave, the fella who was suing him thought he saw George grab the cash box, as he passed through the dance floor door to go downstairs to the exit. The fellow who owned the place followed Geordie and his friends down the stairs, threatening Geordie with a hammer. Bob McGavin, Geordie's university friend who shared an apartment with him in Vancouver, was witness to the attack.

"So," George said, "to protect myself I 'drifted' the guy and stopped him cold."

"So, it was self defence," I said, relieved. "I have done the same thing m'self. Ye can't let somebody get away wi' threatening ye wi' a hammer."

"I really don't think we have to worry, Dad. Shortly after it happened, a police detective came to our apartment to arrest me, but when I explained—and Bob McGavin confirmed my story—the officer decided to investigate further. Later, the police phoned to say they'd dropped the charges. But I guess this guy, hoping we'll fall for his bluff, has decided to sue me privately, himself. That's why Mom got the letter."

With that, Roger Lizee, the owner of the company Geordie worked for in Port Hardy, spoke. "Don't worry about a thing, Dave," he said. "I'll have one of my company lawyers look into it."

I thanked him very much, realizing Roger was doing us a favour. I knew Roger was paying me back for the favour I'd done him. A few years back when he was a young lad getting himself into trouble, I'd taken a chance and given him a job at Colony Farm.

That's where he got to know Geordie, in the first place. At first Roger behaved himself and worked hard at Colony Farm. Later, Roger moved up to Tranquille, where I worked. After awhile though, I'd had to fire Roger. He would party on weekends and not show up for work on the following Monday and Tuesday. I warned him to stop, but he wouldn't listen. Yet, he held no grudge. He knew he deserved to be fired.

In a while, he went to work for himself and he did very well. First, he learned a trade at Kitimat and saved himself enough money to buy several backend loaders. These made him more money. When he sold the backend loaders, he bought a bailiff business in Whalley, then sold that and some property he owned. His profit bought "Hold and Clarke Transportation," in Port Hardy. This was where Geordie worked for him.

By the time Geordie was in university, Roger held the contract to provide transportation over a network of 150 miles of logging roads—all connected to the seven miles of government road leading to the airport at Port Hardy. He transported goods and people between Port McNeill, Port Hardy, Port Alice, and Holberg using freight trucks, Greyhound buses, and Volkswagen vans. He also had other buses, used to transport loggers to and from logging sites and to the airport. In the few years since he'd worked at Tranquille, he'd done very well for himself.

Long before I flew up to Port Hardy to talk to Geordie about his fight, I knew Roger held no grudge against me. Some time back, Roger had written me a letter, thanking me for firing him. In it, he said he realized he'd been wasting his time working for somebody else. Having his own business had helped him to get control of his life.

I flew home and told Peggy, "We don't have to worry, Peggy. Roger Lizee's lawyer will see that fellow suing Geordie doesn't get any of our money. There's no way he's got a case." Sure enough, we never heard or read another word about that fight.

Geordie and some of his friends continued to work for Roger all through their university years. Although the pay was helpful, Roger remained a wild sort and (in hindsight) I don't think his influence on George and the younger fellows was helpful. He taught George how to gamble, and that wasn't good.

The years passed quickly. Peggy continued her volunteer work with the patients. She visited them and enjoyed the feeling of making them happy during the time she spent with them. She enjoyed her associations with the staff at the hospital, and—aside from her annoyance at me for working so hard—she was satisfied with our social life and friends in Coquitlam. She continued to take great pride in her clean house, and also the perfume of her roses when they were in season. Her cooking was always her strength.

Peggy rarely complained, but when she did I knew it was serious. She became worried because her doctor suspected something was wrong with her uterus. After several months of repeated X-rays and tests, she was sent to a specialist. I couldn't believe it when the diagnosis came back. Peggy had cancer of the uterus. At first, I denied the diagnosis, thinking there must be some mistake. It seemed so wrong for Peggy to have this terrible disease.

She needed an operation, but Peggy was fearful of it. She told her doctor of her fear, and he offered her the opportunity to try something new.

"We think there's a good chance it could cure the cancer, without having to go through the operation," the doctor said.

We talked it over and Peggy decided to accept the treatment. We were desperate. As a result, Peggy became one of the first people to be treated by cobalt radiation. The doctors were right. The treatment stopped the cancer, but it didn't end her pain.

Maybe because the treatment was so new they didn't know what they were doing, and so Peggy was never the same. I took over more of the household chores and she slowly regained her ability to move around . . . but from then on she lived with pain in her hip. They had given her too much radiation. They'd burned her. Much of her sweet temperament was lost with the cancer, though I can't say I blame her. Anybody would lose their sweetness with the pain she went through. Christ, I wouldna wish that on anybody!

As Peggy slowly recovered her activity, the increasing time demanded by my family was just one more thing making my life complicated as I got older. I remember thinking back to my father, and to my life as I was growing up in Scotland. These memories were all the more appealing to me now, because—in memory—life seemed so pleasant and simple.

At Colony Farm, my manager's job was anything but that way. Complications were due mainly to increased interference by the unions and the government. With their contracts and their seniority clauses, neither group made my working life easier. Nor, as I saw it, did any of this benefit the farm, the patients, or the animals.

Prior to the unions and government getting into the act, farm managers did the hiring, perhaps with the help of a herdsman, or a foreman, or two. They did the interviewing, they made the decisions about who to hire—based on their own knowledge and their own assessments of the competence of the applicants.

In hiring, I drew on my own experience with farming. I knew most everybody applying for a job said, "Oh sure, I know all about farming. I was raised on a farm," but because I was a farmer myself, I'd ask one or two questions and would soon know if the applicant's statement was a lie. When farm managers did the hiring, it was short, simple, and direct—farmers hiring farmers. This was how Colony Farm became increasingly staffed by good, knowledgeable, dependable men. Knowing how to cull the wheat from the chaff was important. Just like my father before me, I knew my success as farm manager depended upon me hiring men who were good at their jobs.

But when the government and the unions got into it, they made a mess of it. It took them hours longer to administer. It cost more money to achieve poor results. First off, the government hired lawyers to write up rules for contracts with the unions, and that was a damn stupid thing. Under pressure from the opposition party, the government also agreed to seniority clauses, rules about advertising our job openings, and rules about having government representatives do the hiring interviews—though I was still involved.

Problem was, most of the interview questions the government representatives used were made up for hiring people to work in an office. The same tactics were used for hiring stenographers as was used for hiring farm workers! When I wasn't able to be present, people were hired that had never seen a farm, and the hiring committee didn't even realize it! It would make your hair curl, the stuff that went on in their damn system!

For example, one time when I had to be away, they interviewed for a handyman we needed for carpentry work. They hired a man who had been a janitor at Essondale. I suppose they didn't think to check if he knew how to use a hammer, or—from what I learned later—perhaps the people who hired him didn't care.

The first job I gave this fellow was building a fourteen-foot-long water trough, about fourteen inches wide and twelve inches deep. It needed the ends reinforced by strips of metal, mounted on the outside to keep the wood from swelling and opening the joints, allowing the water to leak out. This "carpenter" put the reinforcing strips on the inside, allowing the water to leak through the holes he drilled. He had neither the sense nor the skills to do the job right.

Another time, I asked him to hang a gate. He hung it so it opened into the dirt wall of a dike, maybe less than three feet of an opening for a gate supposed to let a herd of sheep pass through. He also hung the gate so high off the ground, the sheep could have walked underneath, if they'd a mind to.

When I came to inspect the job and saw the way he'd hung the gate, I didn't say anything, at first. Instead, I told him to bring some logs from the pasture beyond the gate.

I went on with what I was doing and a bit later he came back and said, "I can't get the logs through."

"Why not?" I asked.

He looked embarrassed as he replied, "Because I hung the gate wrong."

I told him I'd get the blacksmith to help him out, but—though the blacksmith tried his best—this fella just couldn't catch on. Finally the exasperated blacksmith came back to me and said, "I'm not telling him anything else, he won't listen to me. I can't help him, Dave."

Then I knew we needed to get rid of our so-called handyman, but when I talked to the people who hired him—an executive from the hospital and a government representative—they wouldn't listen to me. This puzzled me until I learned later that this guy was a member of the Freemasons, and so were the people who hired him.

It seems to me there's hardly a Scottish immigrant in Canada who wasn't connected to the Masons in some way. Even myself. In Scotland, my brothers were Freemasons, three of them. You'll never hear anyone admit it, but I know many at Colony Farm got their jobs because of the influence of the Masons.

I said to myself, Dave, with both the Freemasons and the union backing this "handyman," what can you do? The answer, of course was, not a damn thing. I still shake my head when I think about it.

I have to admit though that, with a little conniving, most of the time I managed to get the farm's needs met in spite of government and union interference. I'll give you an example.

One time we needed a tractor man because our man quit. We had to notify Victoria. They advertised the job and informed us it would take about eight months to get to the interviewing stage of the hiring. (I think there was an election on at the time, keeping them busy.) In the meantime, I hired a temporary man, named Al Anderson, on my own initiative. During the eight months of waiting for the government hiring process to grind its way to the interviews, Al proved to be a very good tractor man. As a result, I told him to apply to Victoria for the permanent position, though I couldn't guarantee him the job.

When the time for the interviews came, I looked over the ten applications and noticed there was only Al Anderson and one other man from Saskatchewan who were well-qualified for the job. Because he'd already proved himself, I wanted Al to be hired and a plan came to mind for helping him.

Since I'd farmed on the prairies, I knew prairie equipment didn't usually include a three-point hitch. This hitch lifted the plough, cultivator, or whatever you had at the back of a tractor, so you could turn at the end of a row. (It was a Scottish blacksmith named Ferguson who invented it, by the way). I also knew the government rep at the interview would know nothing about three-point hitches. He'd been born in Ottawa and his father was an RCMP officer. He'd been through university and was smart enough, but he knew very little about the practical side of farming. During the ploughman interviews, I planned to use his lack of knowledge to Al Anderson's advantage, though I told Al nothing of my plan. I made sure Al Anderson was one of the last to be interviewed.

During the Saskatchewan fella's interview, the government rep asked me, "Dave, do you have any questions for this man?"

I said, "Yes. Have you ever had experience using a three-point hitch?"

"No," was the Saskatchewan fella's reply, just as I knew it would be.

"Well," I said, "much of our work involves using a three-point hitch, especially in the seventy-five acres of potatoes here at Colony Farm."

The fellow left the room looking "down in the mouth" because he knew he hadn't made a good impression. The government rep turned to me and said, "What the hell is a three-point hitch!?" Of course, I explained the use of this important piece of equipment to him.

Al Anderson was called in for the last interview, and this time it was the government man who asked, "Have you ever used a three-point hitch?"

"Of course," Al said. "I've been using a three-point hitch for the last eight months while I've been working at Colony Farm."

The government man turned to me with an amazed look on his face saying, "What's this Dave, you've already hired a man?"

"A temporary appointment," I said, letting my annoyance surface. "What the hell do you expect me to do? leave a tractor sitting there idle for eight months while you make up your mind to hire somebody to run it?"

Though he wasn't too pleased with me at that moment, he was a fair man, smart enough to recognize the sense in what I was saying. At the end of the interviews that day, after some discussion, Al Anderson got the job as a ploughman.

I smiled, pleased with myself, but I was also shaking my head at the time, the cost, and the conniving required to achieve what would have been a simple, common-sense hiring decision, under the old system.

Al Anderson became a valued employee for many years. Eventually, because of his experience as a welder in the shipyards during the war, he took over for the blacksmith, Charlie Scott, who retired. We no longer needed a blacksmith who knew how to shoe horses, because all of the working horses were now gone from Colony Farm, just as they were from most farms.

Unlike myself, many Scotsmen are staunch union men. I've never understood this attitude. In 1966, near the end of my career at Colony Farm, the unions called a strike and the men walked off their jobs, leaving only one man with three hundred head of cattle to feed, one man with fifteen hundred pigs, and me responsible for everything else, including the sheep—an impossible situation for both the animals and their keepers.

On the afternoon of the day the strike began, I had finished with the sheep and was heading down the road to help with the feeding of the cattle, over in the barns near Essondale. A group of union men on picket duty were standing at one of the gates. Some were leaning on the gate with the rest peering over their shoulders, gazing into the adjoining field, all having a big laugh. When I got nearer and saw what was amusing them, I was amazed and bloody angry.

A group of big black and brown stray dogs was chasing and chewing the sheep—biting and tearing them apart! Some of the sheep were blood-spattered and helpless, and these union fellas was thinking this was funny. I couldn't understand their reaction at all. More amazing still, I recognized one of the men as one of the shepherds—and a Scotsman at that!

I forced my way through the men and through the gate, with them jeering and laughing as I pushed them out of my way. Then I ran down the field yelling at the dogs as I chased them. I couldn't do much about the injured sheep until I got help, but I sure gave those fellas an earful as I left. They were no longer laughing, at me or the poor sheep, I'll tell ye.

I don't hold much with unions, especially on a farm where people are responsible for animals. Besides, it wasn't me or the animals on the farm that should have been suffering. We didn't set the wages—the government did that.

I don't want to give the impression that everybody who worked at Colony Farm was incompetent. There was a lot more good people than bad. Over the years, the number of good people outweighed the bad, by far. For example, Dr. Quant who is a good friend. He was not only a very good doctor, he was a good man who made sure the patients on parole were being cared for and not overworked. Another young doctor—I can't think of his name right now—used to come down to the farm tennis court, kind and friendly with everybody. An Englishman who headed Public Works was a dedicated man. A gardener named Renton took pride in his work and was interested in the well-being of the place. The head gardener, Joe Hancock—the man in charge of caring for all the flower beds, lawns and greenhouses—had been formerly a gardener at the parliament buildings, in Victoria. He really knew his stuff. The pharmacist was my good friend. Our friendship grew from working together, treating sick or injured animals. He dispensed medicine and pills for the cattle and sheep, and he made sure I understood how the medications were to be properly used. He taught me a great deal about animals and their diseases.

Once we got rid of the dishonest storekeeper, the new store-keeper and his second-in-command—a man by the name of Gilchrist—both were honest and worked hard to keep the supplies ready when we needed them. (Gilchrist originally came from my own Scottish village, and I knew his family well. When he died, he donated his body to medical science.)

I think the only man still living, who was at the farm when I first went there is a good man named Homefelt. He was formerly one of the dairymen who had to milk the cows four times a day, during the early years. Later, he worked in the farm's cannery until he retired. These are just some of the friends and co-workers I remember—part of the good memories I have as farm manager at Colony Farm.

In 1968 when I retired, many of these friends were at the celebration, held outside on Colony Farm's beautiful green lawn. It was organized by Bruce Richardson and his wife, catered by the dietitians, and included most of Colony Farm's workers. Peggy and I enjoyed ourselves, especially laughing at the speeches. I still have the set of lawn furniture they gave us, though many years would pass before I truly took the time to relax on it.

After a two-year search, Peggy and I had acquired a small farm in Aldergrove, which we purchased from our savings for $10,500—cash. Here we planned to move and then raise sheep-dogs, though our plans were relatively short-lived. The familiar old wanderlust returned, prodding me to roam the world as far away as Africa. Like Halley's comet, I guess I still wanted to circle the globe.

CHAPTER FIFTEEN

Aldergrove

We've worn to crazy years thegither;
We'll toyte about wi' ane anither . . .

—Robert Burns, "The Auld Farmer's
New-Year Morning Salutation to
his Auld Mare, Maggi"

When we moved to Aldergrove, the first order of business was to get our new house changed and organized to Peggy's liking. Our son Geordie and his wife Joan came out on weekends to help. Geordie told me he thought of his helping as making up for all the times I'd helped him, over the years. It was a great family time. Peggy loved our wee house when we'd done with it, and both of us felt we'd made a good choice in the place. Neither Peggy nor I missed living within the big institutional environment. Though Peggy's hip was still giving her pain, she was happy and involved. So, after I'd put the house in order, it was time for my interests. I was free to let sheep and sheepdogs became my whole life—for awhile.

You can make a sheepdog do almost anything, if you train it right. I like to work with one dog at a time. Like my father before me, I learned to prefer the females (or bitches) best. Bitches focus on training better than males, then remain anxious to please ye, once trained. My work with sheepdogs took my mind off my pension problems. I put aside thoughts of travelling the globe, because I had no inkling of the opportunity for adventure to come, and because I'd not a lot of money to spend on travel.

My dogs' natural talent for herding gave me great pleasure. Though they're bred to run sheep, without my training they'd run only where they wanted to go, not necessarily where they're needed for the moving of a flock of sheep. With whistles and hand signals, me and the dogs communicated over long distances, understanding one another quickly and clearly.

Though my dogs were well-trained, I was always aware that a dog is still an animal. They can be unpredictable. Even the best dog can get very aggressive in the crowding of a flock of sheep. My guidance made my dogs' efforts purposeful and useful, rather than mischievous or vicious.

I would often forget the passing of time when I was out in the fields with the sheep and dogs. Sometimes I'd forget to come for lunch and Peggy would sure be mad at me. Though I could understand her irritation, because she'd take the time to make me a good lunch, I didn't feel guilty about it. I was never late on purpose.

There were times Peggy'd give me advice and there were times I should o' paid attention. Like the time I accepted the offer of a six-month job from Bruno Jacamazzi . . . the worst move o' my life. I was hired to raise his forty-five thousand chickens. The chicks had to be kept warm, so I had to work in heat and the dampness—so warm I had to change about six times a day. It was six months of my life that felt like six years! Bruno was a good man, but his job was awful. I got out of it as soon as I could.

I've had good dogs and bad dogs in my life, but some stood out in my memory as I worked the fields of Aldergrove. "Nig" was my first dog in Canada. I got him in 1934. Alec Gardner owned him and seven other sheepdogs. Alec farmed sheep beside Stump Lake, in the Okanagan Valley. Whenever I stayed at his place, I'd see him kicking the hell out of Nig, especially when the dog balked at going into his pen, at night. Nobody should treat a dog like that.

During the day, Nig seemed friendly enough and I took a liking to him. I asked Alec if he'd sell him to me.

"Sure, I'll sell him," Alec said, "If you've got the hundred dollars to pay for him."

"A hundred dollars!" I said. "Are ye daft, man? Who's got a hundred dollars during the Depression?"

"Well, that's what I'm asking," he replied. He didn't sound as if he'd ever change his mind. Yet, a week later—when I was ready to leave—he said, "Take the damn dog, if ye still want him. He's no good anyway."

With that, Nig and I left together. I planned to have him accompany me on my trip to buy livestock for our place in Tappen. On the way, my hat blew off. Without waiting for my permission, Nig ran after it and brought it back to me.

"Good dog," I said, smiling.

A little later on, a boy tried to pick up my coat where I had dropped it by the road, as I munched a sandwich. Nig went after him, baring his teeth, growling, and barking. I had to call Nig off. As I sat down to take another bite of my lunch, I thought *Christ, I've got me a good dog, now!*

At Monte Creek, I sat on some grass to have some tea and another bite. Nig suddenly darted across the road, right in front of a car that was moving too fast through the town. The driver hit his brakes with a loud screech and Nig almost made it, but not quite. The tip of the car's bumper caught him, injuring him. I ran over and squatted beside him, stroking his head as he shivered from shock. I could tell it wasn't a fatal injury, but I knew Nig would need time to recover. I needed to get him home, to my brother's Bob's place, but I didn't know how to do that. Just then, a bus driver friend came along, driving his bus. He stopped and I explained.

"Don't worry, Dave," he said. "Give Nig to me and I'll deliver him to Bob."

Bob and Mary nursed Nig and he recovered, becoming an asset to their farm. When I finally left Tappen for good, I didn't have the heart to take Nig away from Bob's family and move him again. He lived happily with them until he was seventeen years old.

One of my favourite dogs, Scotty, came to me from Dr. Woods, a fine man from UBC. He was a friend of Mr. Houstoun's, introduced to me at Houstoun's farm at Hatzic, where he had observed me working the dogs. Dr. Woods had brought Scotty out to Canada and kept him until he was six or eight months old, ready to be trained.

By this time I was one of the directors of the Sheepdog Association, where—for about twelve years—I had helped organize sheepdog trials, bringing sheepdog men together.

Not long before my retirement from Colony Farm, Dr. Woods phoned me. "Dave," he said, "I've got one of the finest bred dogs to come out of Scotland, here. Would you like to train him for me?" Of course, I said yes.

Training Scotty was one of the easiest tasks I'd ever undertaken. He was so wellbred, he was working the sheep after only one day of my teaching him. Though I was working two dogs of my own at the time, neither one had trained as quickly as Scotty.

After I'd had him for about six months and was certain his training was finished, I entered him in a dog trial. I phoned Dr. Woods and told him I'd like him to come and see me run his dog, then he could take him back home with him.

In the field trial, Scotty made me proud. He had those sheep into the pen so fast, the judges couldn't believe it. The crowd of onlooking owners broke into applause, even before the end of the trial. Afterwards, when I came out of the field gate with Scotty, Dr. Woods met me. He had a friend with him and both were all smiles as he shook my hand.

"Congratulations, Dave," he said. "Boy, I've never seen a dog run as good as that! I never thought a man could make a dog that good, that quick! I can't take him away from you now, Dave. You've earned him. He's yours!"

I couldn't believe my good fortune. "Well I thank ye very much, indeed," I said. And with that, I became the surprised owner of the best dog I'd ever trained.

Dr. Woods continued, "I want you to meet a friend of mine, British Columbia's Lieutenant Governor Ross."

As we shook hands, the lieutenant governor smiled and said, "Congratulations on a fine showing. I'm very pleased to see such high standards displayed by sheepdogs showing in Canada. I come from Inverness, in Scotland, so I know good training when I see it. I hope you'll accept this gift from the Sheepdog Association. Keep up the good work."

With that, he handed me twenty dollars, an impressive amount in those days. "Thank ye very much," I said, unable to believe my good fortune. "I promise ye, we'll put it to good use."

Several more times before my retirement, both Dr. Woods and Lieutenant Governor Ross came out to Colony Farm to watch me work with dogs. Mr. Ross was an easy man to get to know. We traded stories, often comparing our lives in Canada to times in the old country.

I learned that he was used to working—like me and most of the Scots I knew. Mr. Ross had come out to Canada first as a bank clerk, then he eventually became the director of the bank. In New Brunswick, he told me he'd been given a piece of land under the "Soldiers Settlement Act," which he vowed he'd never sell, because it was the first piece of land he'd ever owned.

Just before I was ready to retire, Mr. Ross offered me a job, which I wish I could have taken. It was on a farm he owned, on the hill above Campbell River. He wanted me to bring Peggy and come raise his sheep and look after his prize Welsh ponies (slightly bigger animals than Shetlands). By this time, because of his age, he wasn't able to do very much physical work anymore. He showed me a lovely little pond on his farm, which he had stocked with fish.

He said, "If you'll come, I'll even let you fish in my pond, Dave."

As well, in return for my working for him, Ross said he'd welcome me bringing my own dogs and sheep to the farm, and would give me half the profits from the farm's sheep, to keep for myself.

It was a very attractive offer. Peggy and I would have enjoyed the beautiful setting and the job would have been a real help to my retirement, which I needed more than I realized. However, I could only say yes to the offer if Mr. Ross could wait for me to move, until after I retired in August.

I explained, "If I leave Colony Farm before that, I'll have to take a substantial loss to my pension, because I won't have had the required amount of time on the job." Mr. Ross said he understood and agreed to wait, and for the moment I thought things would work out.

Some time later, I informed Mr. Ross I'd discovered my pension had been sadly mismanaged. When I left to go to Hatzic and work for Mr. Houstoun, instead of marking me on leave of absence from my job—which I understood was supposed to happen—the office staff had terminated my pension for that period. This mistake of theirs meant I was going to lose a good amount of money when I retired, unless I made up the difference, as the government was asking me to do.

I told Mr. Ross how angry I was because of this turn of events. Since it was the government's mistake, made by their office staff, I felt it was them who should pay for it. Ross agreed with me, and eventually he was the one who got my pension straightened out, to my own and Peggy's great relief.

It's interesting when I think about it: my training Dr. Wood's dog Scotty not only got me a championship dog, but also allowed me to get to know the lieutenant governor of the province, let me get the help I needed with my pension, and led to a job offer for my retirement days. No matter that two of these opportunities were lost. The first to disappear was the job. In the few months between Mr. Ross's job offer and my retirement, his health became worse. He could no longer manage his little farm. I was never fated to fish in the little pond, or run sheep in those fields. The farm on the hill above Campbell River never became our home.

When Mr. Ross apologized for cancelling his job offer, he explained, "Do you remember my lovely fish pond, Dave?"

"Yes," I said.

"I can't even fish that by myself any more. I can't even put worms on a hook, because of my arthritis. Now, if I fish at all, Dr. Woods or someone else has to attach the bait for me."

"I understand, Mr. Ross," I said. "But I wish I could have worked for you, because I know we share the same appreciation for dogs and Welsh ponies."

My thoughts of retirement years spent winning dog trials with Scotty also came to nought—in as much time as it took me to walk into our house and close the door. Shortly after we won two or three more dog trials, I took Scotty with me to my little place in Aldergrove, while I was still fencing the acreage. I still thought the world of him, felt lucky to own such a well-bred, talented dog.

On this particular day, I'd had the help of my nephew, young Billy McMillan. We'd been working on the fence, talking to each other, not paying much heed to the dog. It started to downpour, so Billy and I made our way back to our house, with Scotty following—or so I thought.

Peggy met us at the door, laughing at the rain dripping down our faces. We went in and sat down for a cup of warm tea and some of Peggy's hot, buttery scones. As I was munching, I suddenly missed Scotty. I went outside in the downpour whistling and calling, but Scotty was nowhere to be found. I looked for him for five days steady.

I asked everybody I knew, "Have ye seen my dog, Scotty?" All to no avail. He was nowhere. It seemed he'd disappeared into thin air. I was devastated.

"Don't worry y'erself," said Peggy, putting her arm around my waist. "He'll come home." But that wasn't to be.

It was much later—too late to prove anything—I heard a rumour that might explain what happened. A neighbour who raised Pomeranians in his kennels was said to have bragged he sold a sheepdog for big money, down in the States. Now, this neighbour never owned anything but Pomeranians, to my knowledge. I think he stole my Scotty, knowing he was valuable.

I only hope the American who bought him gave my dog a good home. Though I eventually had other dogs I was proud of training, none ever gave me the satisfaction I felt when I worked with Scotty at the trials.

Eventually, I was offered an annual job of superintendent of livestock at the Pacific National Exhibition. I was responsible for everything having to do with exhibiting the animals. I made sure the barns and display areas were clean and presentable. I checked animal credentials and tattoos, keeping the owners honest. I supervised the judges. This meant that, for three weeks a year, I worked every day from early morning until late evening. Of course, I didn't do all this work by myself. I had a staff of helpers and my son Geordie came along as my assistant.

Though I was paid for my annual three weeks of working at the PNE, I didn't take this job for the money. I took it for the love of it. These three weeks made sure I'd be in contact with the exhibitors, many of them long-time friends I'd made while working for Tranquille or Colony Farm. In fact, over my years as farm manager, Colony Farm became one of the leading exhibitors at this fair. The breeding program I'd helped build at Colony Farm helped many other PNE exhibitors produce Holsteins, Dorset Horned Sheep, and pigs that met the high standards needed to be eligible for competition. The PNE competition improved livestock overall, for many areas throughout British Columbia. It was a three-week job that was well worth the doing.

I worked as superintendent for the Pacific National Exhibition for two years, before I left on another adventure. Geordie had learned enough about my work at the PNE to take over for me, until I returned. This made me proud of him and happy I'd helped him get the experience. The job helped Geordie because he was newly married and also new to his chosen job, teaching high school. He was glad to have some means of earning money during the summer months, when teachers get no paycheque.

I left the PNE because I was contacted by a fella I'd gotten to know when I worked on the breeding program at Tranquille farm. I think his name was Tommy Willis. He was the man who had been in charge of the federal government experimental farm and knew me because of our involvement together, improving the breeding programme at Tranquille. He was still working for Ottawa, on the range lands outside of Kamloops. He called me and said he remembered my reputation, and what I could do with a farm. He told me the Canadian government could use a man like me in Africa.

"Africa," I said. "Me? I'm seventy years old."

"Yes, Dave," he said. "I know that. I hear they're looking for someone with practical experience to teach farm management on a three-thousand-acre government farm, and the job pays well. You should apply."

I was interested right away, but I told him I'd have to talk it over with Peggy. I said, "I'm willing, but I'll have to ask my wife. I'm not going anywhere without her."

I told her I had a surprise for her to consider. It didn't take her long to make her decision.

She said, "Africa? Oh Dave, I've always wanted to go to Africa. Imagine, at our stage of life, you're offered a job in Africa. Of course I'd love to go."

I said, "Are ye sure? Are ye sure ye're well enough?"

"The last time I went to the doctor, he told me the cancer was cured. The cobalt radiation was successful. I've only the pain in my hip, but that shouldn't keep me from a trip to Africa. Please, Dave. I want to go."

Within a week, we started the process that lead us into one of the happiest, then one of the saddest experiences of our life together.

CHAPTER SIXTEEN

Tanzania

The honest man, though e'er sae poor,
Is king o' men for a' that.
—Robert Burns, "For A'That and A'That"

At first, Tanzania was neither a happy nor a sad adventure. It was frustrating. Peggy and I wondered if we'd made a mistake in taking another job requiring us to deal with the stupidity of big government. The frustration started as soon as I made formal application to the federal government's Canadian International Development Agency for a position as their farming consultant in Africa. After flying to Ottawa, I went through an interview process with three government officials. One of them was a man named Beattie, who headed the African project. At this meeting, I was offered two choices: a job in Basuto (where they hoped to cultivate 135,000 acres of wheat), or a job near Arusha (where they needed someone who knew how to grow corn and beans, as well as wheat). The jobs were on large government farms. They wanted me to teach farm management and how to cultivate crops efficiently.

I'd done my homework to find out about the two locations. I picked the job at Arusha because it would be better for Peggy's health needs. As far as I could tell, in Basuto there was nothing for Peggy's comfort. There were only small villages, away out in the wilds. Arusha—a town of about fifteen or twenty thousand at that time—would be much more comfortable and civilized, or so I thought. Arusha is where I wanted to be, but it didn't happen easily.

Peggy and I were called back for another interview, this time with a very confused woman, who met us at the door of a new building. She said, "Just follow me and I'll take you to my new office," but she couldn't find it. Red in the face, she led us through the hallways looking for it. Along the way, she mentioned a change of plans that would have Peggy and me sent to Basuto.

"What?" I said. "Basuto? We didn't agree to go to Basuto! Why did ye change the plans? Who made these changes?"

Of course, she could tell us nothing, because the papers containing all the details were in her missing office! After about ten minutes of wandering through the new building, she apologized and ended the interview, saying we'd be contacted later.

I thought to myself, *This is a fine thing, to be sent off into Africa by these people. They don't even know where they are . . . and they're in Canada! And they've spent tax payers' money to bring us here to Ottawa!*

This mix-up was almost enough to turn us against going to Tanzania. But we both wanted to see Africa and help its people. We also wanted the chance to enjoy the salary we'd been offered—$20,000 a year, plus $300 a month spending money. I'd never earned this much in my life!

Working for the BC government I wasn't poor, but I never earned over $12,000 a year. Our house was provided free. The free house was a good deal for the government, because it meant I was there for them night and day—twenty-four hours on call. I also got milk a little cheaper—a quart for the price of a pint. That's all, but it was a good deal for us. We grew a big garden of our own, so we needed for nothing. Though Peggy and me weren't rich, we'd always paid cash for what we bought. Never being in debt was important to us and working for the BC Government allowed us to achieve that goal.

But there were problems with my pension. It was small for several reasons. First, I'd never claimed a pension for my army service, because I didn't think I deserved one, having to quit because I was sick, and all. Then, I hadn't had the opportunity to pay into a pension for most of the years I'd been working. The provincial pension plan didn't start until sometime in the 1940s. The plain fact was, Peggy and me needed the money I could earn in Tanzania, so I decided to ignore the woman who couldn't find her own office. It made sense to stick to the plans we'd made in the first interview.

When we had no appointments, Peggy and I wandered about Ottawa and found that city to be the damnedest place we'd ever been! Nobody smiled at us and they all seemed in a rush. I asked a waitress why nobody smiled and she said, "I guess there's nothing to smile about." We were both happy to leave Canada's capital and its gloomy people, and return to BC.

Once home, we finished packing for Africa, happy because my son Geordie and his wife Joan were willing to come and live in our home in Aldergrove until we returned. This set our minds at ease because we knew our home and animals would be tended until our return.

Geordie and Joan drove us to the Vancouver airport. They could see we were both very excited about the trip, especially Peggy. They were all smiles, too. They told us they were very happy for us and we shouldn't worry about a thing. We waved as we entered the passageway to board our plane for London, and then we were on our way.

There were no problems with the travelling, even though it was a long trip. We flew from Vancouver to London, London to Nairobi, then Nairobi to Arusha. We enjoyed the trip, and the best part was meeting the people along the way—fellow travellers from all over the world.

The troubles started when we arrived in Nairobi. Going through customs, Peggy was afraid. It helped somewhat to know the Tanzanian police were modelled after the British, with swagger sticks instead of guns. But suddenly faced with so many Black people and the strangeness—Peggy realized we really were stepping off the plane into the unknown.

"It'll be okay, Peggy," I told her. "Once we get settled on the farm near Arusha, it'll feel more like home."

On arrival, a government official met us, welcomed us, and showed us to our hotel for the night. The next day, another government official—a nice fellow, just doing what he'd been told to do—gave us some disturbing news.

"There's been a change of plans," he said. "Until the paperwork is completed, you'll be waiting with all the other Canadians in accommodations in Dar es Salaam. Then you're to go to Basuto."

"Forget it," I said. "We'll go to Dar es Salaam for as long as it takes to get the paperwork done, but we're not going to Basuto! We're going to Arusha, where our contract says we're to go!"

He said, "But Basuto has the best climate in the world! Other Canadians'll be there, school teachers and others—Donald MacInnes, the mechanic from Manitoba. He and his wife Libby are going there. Basuto's only 180 miles North of here, 6,500 feet above sea level, in a beautiful spot near an extinct volcano—and we need you there for the wheat."

I told him straight, "My wife is better off in Arusha. She has a bad hip and it's too far from civilization in Basuto. If you're not going to place us in Arusha, we're going home. Furthermore—when I get home—I'm going straight to Jack Webster and let him know how you're messing people around."

As soon as I mentioned Jack Webster, the fellow's face paled and he said, "Oh, no . . . for Christ's sake, don't go to that man!" (Jack Webster was a tough, Scottish radio broadcaster in Vancouver, who pulled no punches when reporting on government graft or foolishness.)

As ordered, we joined the other Canadian government employees waiting in Dar es Salaam. Peggy and I were given rooms at the "New Arusha Hotel," first-class accommodations. Later, a French Canadian official came to order us to Basuto, but we paid him no mind, either. There was no retaliation to our refusal, because the government didn't want the publicity I'd promised in response. I learned later that, along with his friend from the forestry service, this French Canadian was too busy chasing women to really care what I did.

The Canadian government officials in Africa soon found out they couldn't mess with me. I was too old and I'd had too many years of experience dealing with government bureaucracies to allow me and Peggy to be pushed around by them.

Next, they wanted us to take temporary accommodations that were substandard in comparison to the New Arusha Hotel, as far as I was concerned. Some of the other Canadian recruits were moved from the hotel into grass huts and other makeshift living arrangements, while they were waiting for completion of placement paperwork, but not Peggy and me.

"We'll stay in the hotel until ye provide the house ye promised us. Thank ye very much!" I said.

While we were waiting, both of us were disturbed by the poverty we saw in the city. It bothered both of us to see the way the African poor had to live. It made us realize how fortunate our own lives had been. We knew our hotel accommodations would look like great luxury to most of the people here.

It took two full months of determined wrangling, until we were finally permitted to go to the farm in Arusha, as we had planned in Ottawa. The two month wait gave us a chance to get to know the other Canadians, also waiting for placement in other locations in Tanzania.

The seven-mile drive to the farm from Arusha was through beautiful African countryside, mostly coffee plantations and other types of farms. Peggy loved the bougainvillea flowers and the lush purple blooms of the many jacaranda trees.

The farm I'd been hired to work on was an impressive joint project of the Canadian and Tanzanian governments. The farmland had been part of a German colony, prior to World War I. At the end of the war, Britain claimed the area. Then an American leased the 25,000 acres from Britain.

The American had made many improvements to the property. He built a fancy cement house and swimming pool, and surrounded both of them with orange trees. He also planted seven to eight hundred acres of coffee and raised imported Herefords. He had a reputation for being typically American, quick on the draw.

I was told he got into trouble a number of times for shooting Masai when they tried to come into his fields. I suppose when his lease was finished, the Tanzanian government reclaimed the rights to the property.

On arrival at the farm, I was introduced to a very smart man named Martin Mengele (pronounced Men-gā-ly), the manager. He was about fifty years old and anyone could see from his appearance he'd looked after himself. He was very fit. Mengele's agricultural specialty had been bananas and coffee. He was an African, a well-educated man who'd worked for fifteen years on another experimental farm under the guidance of a Scottish botanist, a man named Wallace. Mengele was knowledgeable enough to write a book about growing bananas and coffee, but he'd run into trouble when he tried to grow corn, beans, and wheat on this farm. He didn't know anything about them. That's why he needed me.

He said, "Well, Mr. Caldow, I'm pleased you agreed to come to Arusha. I understand you're to be my expert agriculturalist."

"That's the wrong title to give me," I said. "I'm not comfortable being called an "expert" anything. I've had enough experience to know what I'm doing, if that's what you mean, but all my experience was gained in Scotland and Canada. I expect I'll find some things I know nothing about, here in Africa."

Mengele replied, "You know, you're not the first new arrival to say that to me. You'll find others here that'll also tell you how different it is, but your knowledge is welcome and needed here. You'll be the only two Canadians in Arusha, but I think you'll find your other Canadian friends are not too far away for visiting. You must be tired after your trip. Come, I'll show you where you and Mrs. Caldow will be staying. I think you'll like it here."

With that, he took us to a sturdy sixty-foot trailer, built in Kenya—not the house we had been promised, but the trailer looked very comfortable, so we settled in. I soon installed pipes underneath for propane and water. Peggy gave the trailer homey touches, and we were satisfied.

We quickly learned where to go to buy groceries to our liking, in the little nearby town. We found an Asian grocer who stocked smoked cod and haddock from Aberdeen, Scotland, as well as canned sockeye salmon from Russia.

Probably caught by Russian fishermen off the coast of Canada, I thought.

When I asked the grocer where the butcher shop was located, he pointed me in the right direction, then said, "Just follow your nose, mister."

I wondered what he meant, but soon found out. The smell from the meat sold in the open air lead me right to it. Phew, it was strong! I noticed there was worms in the meat, but I'd learned over the years that there's nothing to fear from worms as long as they're well cooked. (Of course, I knew I couldn't let on to Peggy about them worms, or she wouldn't be eating.) All the meat was the same price, so I bought some tenderloin.

Back in the trailer, we cooked it along with some vegetables in our pressure cooker, brought with us from Canada. An Aldergrove veterinarian who had been to Africa had told us to bring the pressure cooker because African meat was often tough (though he said nothing about worms). I thought our supper tasted pretty good, considering. Peggy said it was delicious.

Later, I learned that many of the local children suffered from worms, and I thought it was no wonder that they did, but there was no use complaining. We knew we had to adjust to this strange land.

Peggy invited Martin Mengele for dinner one night, not long after we settled in. She cooked a very tasty meal, as usual, and Mengele loved it, especially the tender meat. He returned often to share dinner with us.

Many of the local children and adults would have been glad of even wormy meat to eat. I heard someone berating one of the farm workers for laziness, but I soon realized the man wasn't lazy at all—he couldn't work for lack of nourishment. You can't expect a man to do a full morning's heavy farm labour when he's only had a banana for breakfast. Nobody could work hard on that!

The locals never wasted food, no matter how it was come by. When four young elephants were chased out of their reserved park area by some older, stronger male elephants, the four of them wandered into one of our cornfields. I watched as two wandered away when they saw our people. Then I noticed the two younger males, determined to stay. They weren't eating our corn, but they were slashing it with their trunks, knocking it down. None of the villagers or farm workers knew what to do, but armed park rangers soon arrived. For most of the day the rangers tried to shoo the elephants out of the corn and back to the park, but those wild animals wouldn't move. Eventually, these two young males had to be shot. That's when I discovered that food was never wasted.

The villagers butchered the elephants, sharing all the meat, taking the food treasure home to be cooked. Nothing was left except guts and tusks. Since the tusks were most valuable, the villagers appointed eight or ten guards—I think they called them "askarees"—to watch over the tusks 'til morning. Even so, next day one of the tusks disappeared. Desperate poor people take desperate risks.

The food we grew was distributed by the local government and sold to the natives. Our corn was ground into meal. Local cooks used it to make cornmeal balls, adding them to meat stews. Special cooking bananas were used the way we use potatoes. Meat dishes were all eaten by hand, because the locals had no utensils.

If they were over fifteen years old, Tanzanian workers were paid only four shillings a day—not enough to buy much food. From ten to fifteen years old, children were paid three shillings a day. No one was paid by the hour. Each labourer was allotted an amount of work to complete before he got his pay. For those with large families, low pay made life very hard. Later, I was surprised to discover that some beggars in town were making more money than farm workers.

Some of our young farm workers were pledged to work for the government for five years, in repayment for their agricultural training. Some had graduated from high school, some had gone to Makere College, some had even been sent to other countries to get a degree in agriculture. They'd studied, but they'd never worked on farms. All they knew was theory.

Mengele wouldn't train them, he said it was up to them to learn. He didn't like them because he thought they were snobby about farm labour. He said they didn't want to work because they'd have to wear work clothes. They only wanted to dress in collars and ties. The result was, they just hung around Mengele's assistant manager, not knowing what to do. I didn't think this was right, but there wasn't much I could do about it, being so new to the farm.

It was hard for both Peggy and me to see the many children in town, plagued with sickness, and so poor they had no clothes at all. There was a law they had to wear clothes, so many of them made do, covered with a threadbare blanket pinned at the shoulder. Kids who couldn't walk had calluses on their hands and knees from crawling since they were babies. They were always hungry, and they'd do anything to get money for food. Five and six year olds would put pepper in their eyes, then go begging with tears running down their cheeks, their eyes red!

There was no such thing as organized sports or summer programmes, such as Geordie grew up with, for Tanzanian kids. Simple basic equipment, like a soccer ball, for example, was unknown luxury. They'd make what they needed from trash. A ball might be rolled up paper or packed leaves tied with string. Young people had to be inventive in this country.

It wasn't long until Peggy got herself involved with the young women at the farm. The girls became interested when they saw Peggy knitting a white baby sweater. About twelve of them, aged fifteen to seventeen, had been through the local agricultural training, but knew nothing about knitting. With Peggy's help, they learned quickly. She made them tea, with scones or pancakes, when they came together.

I think Canada should send people over there to teach these children how to make their lives better. Sending them money doesn't work. Who knows who gets the money? And if they're never taught anything, how can they learn to make improvements?

When Mengele first took me on a tour of the farm, I knew my job here would be a challenge. They did things so differently. I asked the purpose of a huge water hole in the ground. It was attached to a corral by fences that enclosed both the hole and the chute leading down to it. He explained, this was a "dipping hole," built by the former American owner of the farm. The water in the hole had chemicals added, similar to the ones used on sheep, in Scotland. Every Thursday—"dipping day"—the cows had to be rounded up, herded into the corral, then prodded and slid down a mud chute into the water. The resulting dip eliminated ticks. The hole had to be filled with water deep enough to support the cows, cushioning them from injury when they hit bottom.

Good milk cows were special here, rarer than in Canada. This was because most Tanzanians didn't drink as much milk as Canadians. They used milk cows mostly to feed their calves, leaving the cows' milk to disappear as the calves matured.

Later, after Mengele knew me enough to respect my knowledge, he sent me out to buy a milk cow, and I found one with an added bonus. Both Mengele and I were pleased because the cow was pregnant and would soon give us a calf. However, our pleasure soon turned to disappointment.

One Friday morning, Martin Mengele told me the cow had aborted during the night. Puzzled, I talked to some workers about her loss. I discovered she'd been injured the previous dipping day, because the water in the dipping hole was too low to cushion her fall when she went down the chute. When Mengele asked the fella in charge, the excuse was given that a water pipe was broken.

Mengele told him, "Pipes can be fixed, but the loss of a calf is permanent. You're fired, as of today."

A big problem was the lack of machinery needed to do a proper job of preparing the fields for planting. We'd no set of harrows, nor even a roller for making the ground firm enough to support beans. The African ground was so hard and crumbly, I realized we'd have to plant beans very deep, maybe too deep for a good yield, I thought.

Most of the farm fields were filled with weeds about six feet high. I wondered how we would even till them, because a disk cultivator wouldn't work in weeds that big. Much of the land on the farm was wasteland and one of my first tasks was to train the farm workers to reclaim the soil.

In spite of these problems, we managed, mainly because we could laugh at our difficulties. First, Mengele borrowed an industrial-size weed eater from another farm, then he had the tough weeds trimmed to manageable size. Next, I taught a young fella how to use a tractor. When I left him he was driving it pretty good. But once on his own, he got stuck trying to go up a hill. Instead of backing the tractor down to try again, he just kept pushing up, digging the tractor blade deeper by the minute, leaving the back end of the tractor higher with each push. When I saw him next, the red body of the tractor was sitting on high ground, with the wheels spinning in the air! We all laughed at the surprise on the fella's face, as I demonstrated how easy it was to get out of his predicament.

Another time, on another hill, I was teaching a worker how to disk the ground and how to lift the discs for turning. I set him on the Massey Harris tractor, pulling the disks behind. This hill was too steep to allow the tractor to lift the disks out of the ground, while still driving in the same gear. I showed him how to gear down, then I left him to do the job.

Mengele and I came back a while later to check on him. All the time we'd been away, he was moving the tractor successfully up and down the hill. Trouble was, no ground was being broken. He'd lifted the disks out of the ground, but forgotten to put them back down again! He laughed as loud as we did, when he realized what he'd done. It was our sense of humour that pulled us through all the problems towards success.

Once the Tanzanian hills and fields were cleaned of weeds, we made certain they were kept cleared. I trained planters to properly plant, weed, fertilize, and irrigate the dry ground. (They had a hundred ton of fertilizer given to them by Germany, but they'd never known how to use it). From then on, except for the beans, we raised good crops. The Canadian head of the African project (His name was Beattie, I think) couldn't believe we grew eleven-foot-tall corn. He took pictures of it because he said he had never seen corn that tall in that part of Africa. The wheat also was successful. We harvested seventy-two bushels an acre, while the average yield in Saskatchewan was twenty-five bushels per acre, at that time.

However, Mengele and I were not proud of the beans. We found no way to create ideal conditions for their growth. We did our best to prepare the ground, but the hard, crumbly soil defeated us. As the little green bean shoots fought their way to the surface, I trained children to keep them weed free with hand hoes. Trouble was, when they hoed, they often killed the beans as well as the weeds. The surviving plants were spindly, producing a poor crop at best. Our one failure.

Whenever we had free time, Peggy and I would meet with our Canadians friends we knew from Dar es Salaam, then we'd tour the countryside together. Dora and Bill Davis, who had worked at Aggasiz experimenting with different types of grasses, sometimes came with us. Probably most often, Libby and Donald MacInnes, from Manitoba, were along.

We became good friends. Donald was a mechanic, brought to Basuto to teach the maintenance of tractors and combines, needed to improve African wheat cultivation. He was one of the finest men I'd ever known, with a temperament like that old movie star, Will Rogers. Donald talked slowly, considering his words carefully before he spoke them. If he disagreed with a person, he would never call him a liar. He'd just talk the disagreement through. Many times, I wished I was more like Donald MacInnes. We became good friends, even though we worked some distance apart. We visited them in Basuto, where the farm was cultivating 45,000 acres of wheat. On the drive from Arusha to Basuto, both Peggy and me were impressed with the strange-looking Boabab trees.

We made many trips. We travelled to Fort Jesus at Mombasa, on the coast. On the way we took pictures of Mount Kilimanjaro.

It was so hot at Fort Jesus, Libby couldn't get enough water to drink. We also drank beer, but still, Libby said she had never been so thirsty. Fort Jesus was impressive and also haunting. There was a pit inside this fort, where they let people starve to death during the slave trade. One of Peggy's favourite places was Lyamangu, an old coffee plantation we visited a number of times where we took pictures of the African women sorting the coffee. I enjoyed the racetrack in Nairobi, Kenya.

We invited Donald and Libby to go with us to see Africa's biggest wild game park—located at the bottom of an extinct volcano crater—but Libby had no interest in going. She'd heard about the steep road down the side of the crater. She'd heard about a busload of tourists killed there a year before, and decided she wanted no part of it. I was also concerned because I have always had a fear of heights, but I'd heard sights in the park were worth the risk. Besides, Peggy really wanted to go.

I remember the park is in the shadow of Mount Kilimanjaro—I think it was called the "Tsabo Game Park." To be safer, we were required by law to hire African park guides. Two of them drove our group of six tourists, in two jeeps. The jeeps had light canvas covers, but otherwise were open for good viewing. As a result of the open sides, my fear of height on the trip into the volcano was my worst nightmare come true, or so I thought, until we travelled out.

Our jeeps moved slowly along the narrow dirt road, edging the steep drop-offs. The sight of countless wild animals moving slowly across the Serengeti Plains eventually eased my fear. But I felt breathless and my hands ached from gripping the sides of the jeep.

Finally, when we reached the bottom of the cliffs, we became just like the herds of animals, moving slowly across the Serengeti, ourselves. Our guides knew their business, moving to about ten feet from five healthy-looking, tawny-coloured lions.

Our guide said, "It's their breeding season and—from the looks of them—they're not hungry. They're not interested in us. We're lucky, today."

I sure hoped he was right about being lucky, though I'd no sense the lions were threatening. They were lying around just like satisfied house cats—very big satisfied house cats. Peggy was excited, really enjoyed observing them.

A little further on, we noticed a rhinoceros lying by the side of the trail, off a bit in the grass. At first he looked dead, he was so motionless. We stopped and watched him. Then we noticed his belly was moving in and out slowly. He was breathing, just asleep.

I said to the guide, "That rhino looks so peaceful, as if I could pat him."

The guide's white teeth flashed as he laughed. "He'd have you before you could get back to the jeep!" Needless to say, Peggy was relieved I stayed where I was.

On the flat Serengeti with the animals, it was exciting but none of us felt much fear. It was more curiosity and pleasure we felt—all of us happy to see the many animals we'd only seen in picture books, or movies. The real fear happened when we left the plains and made the return trip up the narrow road on the side of the crater.

Close to the base of the cliff, our two jeeps suddenly were confronted by a convertible with four men in it. The convertible had to stop, because our guides used our jeeps to block the narrow road.

"What's happening?" we were all asking.

Our guide muttered, "They're attempting to enter the park without guides. But we've got 'em."

What now? I thought.

Each guide jumped out of his jeep, walking towards the convertible. It was a tense moment, but luckily, the lawbreakers weren't armed.

"You're under arrest! Turn around," the tour guide ordered the car driver. "Move your car behind this jeep," he said motioning to the jeep where we were riding. "I'll move my jeep in behind your car, then we can both keep an eye on you."

Keeping an eye on them was what terrified me. As we inched and bumped our way higher and higher up the side of the cliff, our guide kept taking his eyes off the road to look back at the convertible. I didn't say much, but Peggy could see I was scared as hell. I was white as a sheet! I clenched my fists so hard, my nails cut into my palms! By the time we reached the top of the cliff and stopped to stretch our legs, I was frozen with fear. It was only Peggy's laughter that loosened me up. She couldn't understand why I was so afraid, knowing I'd faced much worse in my lifetime. She laughed at me, then I started laughing, and everybody else laughed with us.

Laughter and Africa go together in my memories, in spite of the challenges we faced during our two years there. Peggy was so happy, until her hip became too painful. We thought it was the cancer had returned, so she went to a German doctor, who was doing research growing cancer cells. He told Peggy she should return to Canada, because he knew of no doctors in Africa who could help her.

Peggy said, "I've no desire to return to Canada this soon, Doctor. Dave and I are enjoying ourselves here."

But the doctor was firm. "There's no way for you to get help, here," he said.

Because I knew the doctor was right, I regretfully informed Martin Mengele we'd be leaving.

"I want you to know, we're leaving for Peggy's good," I said. "We'd both prefer to stay. The two years I've been working here have been the happiest years of my life. I've felt my work's been very appreciated, and I've never been sick a day, since we arrived. Not even a cough."

"I'll miss the both of you," Mengele replied. Yet I had a feeling there was something he wasn't telling me. A few days later, he walked over to me, as I stood looking at one of the cows.

"Dave, I've made a decision," he said. "I'm going to retire, too. I've had a successful two years working with you and I want to retire on a 'high note.' I'll be leaving the same time as you." His news surprised me—as it did many others—but it led to one of the biggest retirement and goodbye parties I've ever seen.

Peggy made me dress for it, the fanciest I'd had to dress since coming to Tanzania. She told me to wear my best suit, then checked my tie to be sure it was straight.

"After all, ye must look good when the president of Tanzania is sending a representative to shake your hand," she said, smiling.

The party was held outdoors, on the farm grounds, not far from the front of our trailer home. I estimated there were between two and three hundred people at the party, Tanzanians and all the Canadians in Tanzania. When we stepped out on the steps of the trailer, we couldn't believe the number of people. The huge crowd consumed twenty-four cases of wine (twenty-four bottles to a case) and two drums of "pombee," (forty-five gallons to a drum). Pombee (a Swahili word) is a potent homebrew. It's more like a meal than a drink, made from bananas and thick as a milk-shake. Very filling. If you have too much of it, you'll soon be flat on your back looking up at the sky, not knowing where you are. I had enough to enjoy myself, but not enough to spoil a good party. Both Peggy and I enjoyed the celebration, the speeches, the music, the laughter. It was a great way to say goodbye to all our Tanzanian and Canadian friends.

Both Peggy and I were sorry about leaving. We knew we'd be missing all the new experiences and the countryside, but more than anything, we knew we'd miss the people. It was for the best that we said our goodbyes during the festivities, then left quietly, two days later, for Arusha and the beginning of our trip home to British Columbia.

EPILOGUE

An monie a sair daurk we twa hae wrought,
An' wi' the weary warl' fought!
— Robert Burns, "New Year Morning Salutation"

Close to ninety years have passed since I was a young lad, running in the fields of Scotland, celebrating Halley's Comet. These days, the vast fields of my labour are gone from my life. I now work in a small garden plot at the back of my seniors' care home, yet even this bit of land is bountiful, adding joy to my living. I'm still called "the farmer," and friends smile when they see my tomatoes reddening and growing juicy on the vines.

In my ninety-eight years, I've travelled to places beyond my dreaming when I was a boy. Many friends have brought me joy and fill my mind with good memories. I've a warm family that cares for me, but I do miss Peggy. She was a brave woman and good wife to me.

After we returned from Africa, on our fiftieth wedding anniversary, Peggy and I shared a joyful cruise to Alaska, a close time and happy time for us. Not long after this—because of Peggy's need for doctoring—we had to sell our wee home in Aldergrove. The long drive to doctors' offices in town made her very uncomfortable from the pain in her hip, so we moved to a more convenient apartment in White Rock, where we lived for several years. It was here Peggy died of a stroke, lasting only a few days after she was hit with it.

There was no funeral, because I did what she'd told me to do. Travelling alone, I took her ashes to Paul Lake, carrying them in a small, foil-covered box. Some I spread on the ground around the house where we'd stayed when she was expecting her daughter, Ethel. Some I placed around the house where we'd stayed on weekends away from Tanquille. Some I spread by the road at the lakeside, where she'd often walked. By doing this, I returned her to a place she thought most peaceful, a place where we'd both found happiness many times.

It rained on the day I put her to rest. Though the rain blended with my tears, there was satisfaction in doing what she'd asked me to do. Peggy had endured so much for so long, it was a relief to know her suffering had ended.

The world continues to change. The Provincial Mental Hospital (Essondale) is no more. Colony Farm fields are overgrown, with much of the land returned to serving the needs of herons and other birds, to the small wild animals and plants that have grown there always. The house Peggy and I shared when I was farm manager is no longer used. Peggy's roses and gardens are overgrown. I understand the community is making plans, considering the Colony Farm lands for a mix of changes, perhaps a nature park, perhaps housing development, perhaps other things.

I miss the way it was, yet change is nature's way. Nothing ever stays the same. Even Halley's Comet shone dimmer on its last visit. (Perhaps you could say I'm becoming a little dimmer too, because my memory's sure not a shining talent anymore.)

Yet—like me—that comet's still travelling, still circling back to the skies of Scotland, as I've done in aeroplanes many times over the years, and still do in my mind. Though I'm no longer confident enough to travel physically that far, I know my thoughts will forever return to my father, to his good Scottish teaching, to the family we shared.

I have few regrets about the life I've lived. I wish I had been a better father and a better husband, but I did what I could, considering what I'd known. I always put my job first, but I wish now I'd not done that so much.

Yet in spite of my not spending enough time with them, the children I helped raise have become very successful adults.

Geordie became a teacher and married his high-school sweetheart, Joan Sargent. He was a high-school principal and now is retired from being Director of Personnel for the Coquitlam School District. He and Joan—a teacher too—have been married for over thirty years. They gave me my grandsons, Marty and Curtis. My nephew, Bobby Caldow, went on to university, married happily, and became very successful, selling large farming machinery. He's now retired in Kelowna and we still visit often. My niece Jeannie is almost a daughter to me. Her husband Gino is my good friend. Ethel (Peggy's daughter) turned out just fine and to me she looks the living image of her mother, more so every time I see her.

If I were giving advice to my grandchildren, who're not farmers, I'd say pretty much what my father said to me: Work hard at a job ye enjoy. Give it ye'r best and be honest. But I'd add one more thing. I'd tell them to take care o' their friends, because friends'll help ye, even when ye don't know they're doing it. But I think my grandsons'll both turn out just fine, with or without my advice.

For certain, I know the standards of British Columbia farming practices and the quality of the livestock were improved because of the work done by the Farmers' Institute, Tranquille, and Colony Farm, and I've had a hand in that. In provincial, national, and international competitions, over the years we were recognized as producers of excellence. I've worked beside and been friends with some of the best men in agriculture, and I made sure that most of the time I worked at what I enjoyed. Farming gave me a good living. Farming took me travelling to the other side of the world and back again. Farming made me a farm manager, just like my father.

All in all, I'd say I've accomplished the dreams of the wee lad who ran long ago in the fields of Scotland.

NOTES

Prologue

1 Carl Sagan and Ann Druyan, *Comet* (Toronto: Ramdom House, 1985), pp. 122-23.

David mentioned you could see his comet in daylight, so it probably wasn't Halley's; as Sagan and Druyan have noted: "About once in a human lifetime, on average, a comet appears that can be seen in the daytime sky, even very close to the sun. The Great Comet of 1910, which in people's memories is sometimes confused with Halley's Comet which arrived later that year, was such, and was called the Great Daylight Comet."

If this were a fictional work, a writer could play on the ironic possibilities of this misidentification. However, since David's interpretation of reality forms the basis for his decision making, this symbol retains the meaning he gave it. David's reaction to this event is much more telling than any astronomical facts. One can imagine how different his life would have been if David had remembered cowering in fear, or if he had remembered the comet as shining solely on him. Instead, he remembered working with others to achieve success and approval from both siblings and the two adult role models most important to him: his father, a very powerful, celebrating man and his mother, a strong, supportive woman, both undaunted by life's unknowns.

Chapter 3

1 John Ferguson Snell, *Macdonald College of McGill University: A History from 1904–1955* (Montreal: McGill University Press, 1966), p. 76. All other references to Macdonald College are confirmed by this publication.

Chapter 4

1 Anita Wisti and Jim Airhart, *Glimpses of Marmora* (Marmora: self-published, 1974) p. 59.

2 R.G. Walsh, "Deloro." Speech to the Marmora Chamber of Commerce, 2 April 1953 (Village of Deloro archive, Marmora Historical Society), p. 2.

3 Ibid., p. 4.
4 Wisti and Airhart, *Glimpses of Marmora*, p. 61.
5 Walsh, "Deloro," p. 4.
6 "Deloro Village Refuses to Die," *Belleville Ontario Intelligencer*, 7 November 1962, sec. 2, p. 15.
7 Various articles, *The Deloro Once-A-Week*, 1, 6 (24 April 1920).
8 Donald M. Liddell, ed., *Handbook of Non-Ferrous Metallurgy* (New York: McGraw Hill, 1926), p. 811.
9 Ibid., p. 812.
10 Ibid., p. 814.
11 *The Safety Committee of Deloro Smelting & Refining Company Ltd. Its History, Organisation, Constitution, Etc.*, pamphlet, Village of Deloro.

Chapter 5

1 Details from an interview with John Thompson, at Claresholm Museum, July 1994.
2 Claresholm History Book Club, *Where the Wheatlands Meet the Prairies* (Claresholm, AB: 1974), p. 298. Subsequent references to Claresholm families were confirmed by this book, pp. 208, 345, 483.

Chapter 6

1 Salmon Arm Museum and Heritage Association *A Salmon Arm Scrapbook* (Salmon Arm, BC: Salmon Arm Museum and Heritage Association, 1980), pp. 47-51.

Chapter 7

1 Ken Favrholdt and John Stewart, "Adult Education for Farmers," *Kamloops Daily News*, 25 November 1988, p. 24.
2 Deleeuw family, *Kamloops Sentinel* index (cards), Kamloops Museum and Archives.
3 Favrholdt and Stewart, *Kamloops Daily News*, 25 November 1988, p. 24.
4 Jimmy McPhee, "Pete Moore Has Breeding on the Brain," *British Columbia Weekly*, 15 March 1948. (Caldow family saved clipping, page number not saved.)
5 Pierre Berton, *The Great Depression: 1929-1939* (Toronto: McClelland and Stewart, 1990), p. 350.

Chapter 8

1 "The Insane Asylum," *Victoria Colonist*, 14 December 1896 (microfilm, Provincial Archives, Victoria, BC).
2 "Return to an Order of the House for all papers and correspondence relating to the appointment of a Resident Physician for the Asylum for the Insane at New Westminster," *Province of British Columbia Sessional Papers—3rd sess., 4th parliament*," 1885, pp. 329-30 (Provincial Library, Victoria, BC).
3 Horace A. Sheridan-Bickers, "The Treatment of the Insane: Farming as a Cure for Madness, British Columbia's Novel Experiment," *Man to Man*

6, no. 12 (December 1910): 1050-59 (BC Archives Library Catalogue, Victoria).

4 "Annual Report on the Asylum for the Insane," Province of British Columbia Sessional Papers, vol. 2, 3rd sess., 19th parliament, 1939, p. xii (Provincial Library, Victoria, BC).

5 Pierre Berton, *The Great Depression:1929-1939* (Toronto: McClelland and Stewart, 1990), p. 373.

6 Ibid., p. 374.

Chapter 9

1 Pierre Berton, *The Great Depression: 1929-1939* (Toronto: McClelland and Stewart, 1990), pp. 502-507.

Chapter 12

1 Jimmy McFee, "Pete Moore Has Breeding on the Brain," *British Columbian Weekly*, 15 March 1948, p. 27.

Chapter 13

1 Sybil A. MacFarlane, "Tranquille: Memories of the 1940s," *British Columbia Historical News* (Winter 1991-92): 24-25. Subsequent references to Tranquille were confirmed by this source.

Chapter 14

1 Handwritten notes, "Crease Clinic," (first page unpaginated), microfilm, Provincial Archives, Victoria, BC.

2 Ibid., third page.

3 "Annual Report on the British Columbia Mental Hospitals–1932" Province of British Columbia Sessional Papers, 5th sess., 17th parliament," 1933 (Provincial Library, Victoria, BC) p. 11.

4 "Annual Report on the British Columbia Mental Hospitals–1948" Province of British Columbia Sessional Papers, 5th sess., 21st parliament, 1949 (Provincial Library, Victoria, BC), p. 50.

5 Stan Burke, "Bossie is Emancipated; She'll Have Fine Parlour," *British Columbia Weekly*, 15 March 1948, p.28.

6 Jimmy McPhee, "Pete Moore Has Breeding on the Brain," *British Columbia Weekly*, 15 March 1948, p. 27.

Books in the Life Writing Series Published by Wilfrid Laurier University Press